A Text Book Of

MAINTENANCE AND REPAIRS OF STRUCTURES

(22602)

Semester - VI

FOR

THIRD YEAR DIPLOMA COURSE IN CIVIL ENGINEERING GROUP

As Per MSBTE's 'I' Scheme Syllabus

SALIL P. DESHPANDE

B. E. (Civil)
H.O.D. Civil Engineering Department
V. B. V. Polytechnic,
Vasai Road (W).

AKSHAY P. JOSHI

B. E. (Civil)
Lecturer, Applied Mechanics Department,
Government Polytechnic,
Jalgaon

N4554

MAINTENANCE AND REPAIRS OF STRUCTURES ISBN 978-93-89825-42-8

First Edition	:	**January 2020**
©	:	**Authors**

The text of this publication, or any part thereof, should not be reproduced or transmitted in any form or stored in any computer storage system or device for distribution including photocopy, recording, taping or information retrieval system or reproduced on any disc, tape, perforated media or other information storage device etc., without the written permission of Authors with whom the rights are reserved. Breach of this condition is liable for legal action.

Every effort has been made to avoid errors or omissions in this publication. In spite of this, errors may have crept in. Any mistake, error or discrepancy so noted and shall be brought to our notice shall be taken care of in the next edition. It is notified that neither the publisher nor the authors or seller shall be responsible for any damage or loss of action to any one, of any kind, in any manner, therefrom.

Published By :

NIRALI PRAKASHAN

Abhyudaya Pragati, 1312, Shivaji Nagar,

Off J.M. Road, PUNE – 411005

Tel - (020) 25512336/37/39, Fax - (020) 25511379

Email : niralipune@pragationline.com

➢ **DISTRIBUTION CENTRES**

PUNE

Nirali Prakashan : 119, Budhwar Peth, Jogeshwari Mandir Lane, Pune 411002, Maharashtra

(For orders within Pune) Tel : (020) 2445 2044, Mobile : 9657703145

Email : niralilocal@pragationline.com

Nirali Prakashan : S. No. 28/27, Dhayari, Near Asian College Pune 411041

(For orders outside Pune) Tel : (020) 24690204 Fax : (020) 24690316; Mobile : 9657703143

Email : bookorder@pragationline.com

MUMBAI

Nirali Prakashan : 385, S.V.P. Road, Rasdhara Co-op. Hsg. Society Ltd.,

Girgaum, Mumbai 400004, Maharashtra; Mobile : 9320129587

Tel : (022) 2385 6339 / 2386 9976, Fax : (022) 2386 9976

Email : niralimumbai@pragationline.com

➢ **DISTRIBUTION BRANCHES**

JALGAON

Nirali Prakashan : 34, V. V. Golani Market, Navi Peth, Jalgaon 425001, Maharashtra,

Tel : (0257) 222 0395, Mob : 94234 91860; Email : niralijalgaon@pragationline.com

KOLHAPUR

Nirali Prakashan : New Mahadvar Road, Kedar Plaza, 1st Floor Opp. IDBI Bank, Kolhapur 416 012

Maharashtra. Mob : 9850046155; Email : niralikolhapur@pragationline.com

NAGPUR

Nirali Prakashan : Above Maratha Mandir, Shop No. 3, First Floor,

Rani Jhanshi Square, Sitabuldi, Nagpur 440012, Maharashtra

Tel : (0712) 254 7129; Email : niralinagpur@pragationline.com

DELHI

Nirali Prakashan : 4593/15, Basement, Agarwal Lane, Ansari Road, Daryaganj

Near Times of India Building, New Delhi 110002 Mob : 08505972553

Email : niralidelhi@pragationline.com

BENGALURU

Nirali Prakashan : Maitri Ground Floor, Jaya Apartments, No. 99, 6th Cross, 6th Main,

Malleswaram, Bengaluru 560003, Karnataka; Mob : 9449043034

Email: niralibangalore@pragationline.com

Other Branches : Hyderabad, Chennai

niralipune@pragationline.com | www.pragationline.com

Also find us on [f] www.facebook.com/niralibooks

Note : Every possible effort has been made to avoid errors or omissions in this book. In spite this, errors may have crept in. Any type of error or mistake so noted, and shall be brought to our notice, shall be taken care of in the next edition. It is notified that neither the publisher, nor the author or book seller shall be responsible for any damage or loss of action to any one of any kind, in any manner, therefrom. The reader must cross check all the facts and contents with original Government notification or publications.

This book is dedicated to my Father

Late Shri. Prakash Balkrushna Deshpande

Author
Er. Salil P. Deshpande
salil_pd@yahoo.com

...

This book is dedicated to my Parents

Shri. Pramod Shrikrushna Joshi and
Sau. Sunita Pramod Joshi

Author
Akshay P. Joshi
akshay.joshi1993@gmail.com

Preface ...

A.P.Joshi

This Text Book of **"Maintenance and Repairs of Structures"** is strictly written as per the new syllabus of 'I' Scheme effective from June 18, prescribed by the Board of Technical Examinations, Mumbai, Government of Maharashtra. The present book of "Maintenance and Repairs of Structures" is in pursuance of the urgent need of Civil Engineering students. This will enhance the knowledge of students as per need of industry.

"Maintenance and Repairs of Structures" is a specialized subject of Civil Engineering and has opened various new career options for civil engineer. It involves for execution of works, sanctioning of jobs, preparing estimates for Repair of buildings, dams, bridges and other civil structures.

The arrangement of the topics in the book is made in order to ensure a smooth flow of the subject for better understanding of the concepts. Alongwith theory, as subject is Practical based, details procedures and photographs of field work are also given in this book as per the need of the students. So we hope that all need for this subject will be completely fulfilled in this book.

Our sincere thanks are to Shri. Dineshbhai Furia, Shri. Jignesh Furia, Shri. Pradeepbhai Furia and whole Staff of Nirali Prakashan, especially Mr. Shashikant Patel and Mr. Jayant Dedhia. We are also thankful to Mr. Santosh Bare (D.T.P.), Mrs. Manasi Pingle (Proof Reading), Miss. Chaitali Takale (Fig. Drawing) from Pune Office for making this effort possible.

We take this opportunity to express our gratitude to our Principal, HOD and all staff members of our college for their encouragement and appreciation and also Special thanks to our family members and well wishers for their patience and encouragement.

We express our indebtedness to all books, reference, periodicals and journals, which helped in the preparation of this book.

Although every care has been taken to check mistakes, yet it is difficult to claim the perfection. Any errors, omissions and suggestion come forward from teachers, students and field engineers to improve the subject matter will be thankfully acknowledged and incorporated.

S. P.Deshpande
A.P.Joshi

4.4   Repair techniques : Grouting, patch spalling replacement or delaminating and epoxy bonded mortar.

4.5   Repairing methods for minor and medium cracks include epoxy injection, grooving and sealing, shotcrete, stitching, grouting and guniting.

4.6   Repairing methods for major cracks (width more than 5 mm) include fixing mesh across cracks, dowel bars, RCC band and installing ferro-cement plates at corners and propping.

4.7   Effects of dampness in wall, damping repair techniques such as replacement or inserting DPC in brick wall, bituminous painting, painting using water proof solution and cement with adhesive gum.

4.8   Causes and remedies of foundation settlement, improvement techniques by compaction, intruding sand piles, stone columns and grouting cement slurry.

5. MAINTENANCE AND REPAIR METHODS FOR RCC   (M-14, H-14)

5.1   Probable location of cracks in RCC elements, various causes of RCC failure.

5.2   Causes of dampness in roof slab and its repair techniques such as mud phuska with brick tile topping, lime concrete terracing, ferro-cement topping and brick coba.

5.3   Repair methods for cracks in RCC structures such as epoxy injection, grooving and sealing, stitching, rebaring, grouting, spalling replacement, jacketing, shotcrete and gunitting.

5.4   Repair of corroded RCC element : exposing and undercutting rebar, cleaning reinforcing steel, compensating reinforcement and protective coatings.

5.5   Repair methods for honeycomb and larger voids.

6. STRUCTURAL AUDIT AND BUDGET   (M-12, H-12)

6.1   Necessity and importance of structural audit and budget estimation.

6.2   Distress survey, detailed inspection, recommendations for budget estimation.

6.3   Steps involved in structural audit and budget estimation.

6.4   Format preparation for structural audit including general information of building, building data, complain reported by users, inspection of internal and external areas of building.

6.5   Overview on rules and regulations of structural audit and budget estimation as recommended by competent authority such as Public Work Department.

•••

Contents

•••

Chapter **1**

BASICS OF MAINTENANCE AND REPAIRS

Weightage of Marks = 08, Teaching Hours = 06

Syllabus

1.1　Maintenance and its classifications, repair, retrofitting, re-strengthening, rehabilitation and restoration.

1.2.　Necessity, objectives and importance of maintenance and repairs.

1.3　Factors influencing the maintenance and repairs.

1.4　Advantages and limitations of maintenance and repairs.

1.5　Approach of effective management for maintenance and repairs.

1.6　Periodical maintenance, maintenance manual containing building plan, reinforcement details, material sources, maintenance frequency, Pre and post monsoon maintenance.

Objectives

After learning this chapter, student will be able to –

- Explain the necessity of maintenance of the given civil structure.

- Know the factors that influence on maintenance of the given structure with justification.

- Explain the concept of retrofitting, re-strengthening, rehabilitation, and restoration.

- Know the periodical maintenance and its manual, monsoon maintenance, maintenance history sheet.

INTRODUCTION

Definitions :

- **Defects :** The defects are defined as, "the flaws those creeps into structure because of design mistakes or poor workmanship during construction before the service life begins". The flaw which has a potential to lead to a failure later becomes a defect.

- **Distress :** Distress is "a term used for cracks, pop-outs, decay or corrosion in the structure". It can be considered as the indicator of the defects present or damage.

- **Deterioration :** It is "the continuous loss of the material properties due to various degradation factors". Unlike defects it may not surface at the beginning of the service life of a structure, but is time-dependent. Some forms of deterioration may develop early in the service life of structure also.

- **Repair :** It is "the process of making something that is damaged or deteriorated or broken, to good condition". It may be the modification of a damaged structure, partially or fully. It includes reconstruction of non-structural walls, repairing of cracks, checking and repairing electrical connections, plumbing, ventilation etc. It does not cover the strength aspect of the structure.

- **Renovation :** "Process of substantial repair or alteration that extends a building's useful and serviceable life" is called renovation.

- **Remodeling :** "When Renovation is done for changing the utility" is called remodeling.
- **Restoration :** Restoration is defined as, "the process of re-establishing the materials, form and appearance of a structure". The restoration is performed to attain the strength of existing building to the original design strength. The restoration enables to get at least the original strength of structural members. The restoration can be done by grouting, strengthening using wire mesh or rebuilding the cracked portions using rich non-shrinkable mortar etc.
- **Strengthening :** It is defined as, "improving the structure to a level of strength, higher than that initially designed by modifying the structural member which may not necessarily be damaged structure". It is the process of increasing the load-resistance capacity of a structure or structural member.
- **Rehabilitation :** It is "the process of strengthening a building or an area to its, previous conditions". Rehabilitation is "the process of restoring the structure to service level once it had". The rehabilitation of structurally deteriorated RC structures becomes necessary since the structural member ceases to provide required strength and serviceability.
- **Retrofitting :** "Assessing the existing condition of the structure and deciding which component of the structure should be repaired or restored based on all the future requirements of structure" is called retrofitting. The retrofit enables to increase the original strength of the building. The Retrofitting includes addition of shear wall of diagonal braces, modification of roofs, strengthening of foundation, modification of building plan etc. It is the process of strengthening of structure along with the structural system, to comply all relevant and recent codal provisions in force during that period.
- **Demolition :** "The process of pulling down of the structure not deemed to be fit for service" is known as demolition of structure.
- **Maintenance :** Maintenance is "the act of keeping the structure in working condition by inspecting and repairing it regularly".

1.1 MAINTENANCE OF CIVIL ENGINEERING STRUCTURES

- The proverb "A Stitch in Time Saves Nine" is applicable to Civil Engineering structures. It means that a repair if attended urgently would avoid a costly one at a later stage.
- Maintenance is generally preventive in nature. It includes inspection of structure and surroundings and repair works necessary to fulfill the intended function or to retain original standard of service.
- Any building when built has certain objectives and during its total economic life, it has to be maintained. It is interwoven with good quality of housekeeping. It is mainly governed by the quality of original material and workmanship of construction. The owners, engineers, tenants and the maintenance agency are all deeply involved in it and also share a responsibility. Situation in which all these agencies merge into one is ideal and most satisfactory.
- The two processes considered are, the work carried out after inspection in anticipation of failure and other the repair work taken up after failure. The former called as **preventive maintenance** and the latter as **corrective maintenance**. The prime objective of maintenance is to maintain the usability and performance of the structure and its services to provide an acceptable and efficient operating environment to its users.
- The responsibilities for maintaining and managing buildings in safe and sanitary conditions rest with owners. A proper management system and timely maintenance of buildings prevent its deterioration, keeping it safe and provide a pleasing and comfortable living environment and also uphold its value.
- Surveillance may be combined with inspection of structure for maintenance. It would be beneficial to occupants in engaging the same technical team in carrying out both duties.
- Defects in structure create hazards leading to disastrous injuries. Most defects can be determined through visible or detectable symptoms at their early stages. If not timely rectified, these minor defects can develop into fatal ones. It may further cause failure or sudden collapse of structure, endangering lives or becoming more costly to rectify.

- Inadequate or inappropriate repair will also result in frequent failures, causing trouble to the occupants and public. It would also accelerate the process of depreciation of assets.

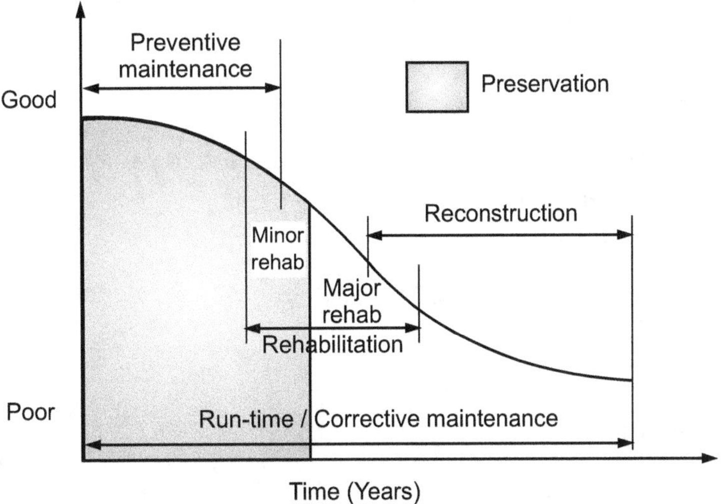

Fig. 1.1 : Times based structural health graph

1.2 TYPES OF MAINTENANCE

The Maintenance of structures may be classified as,

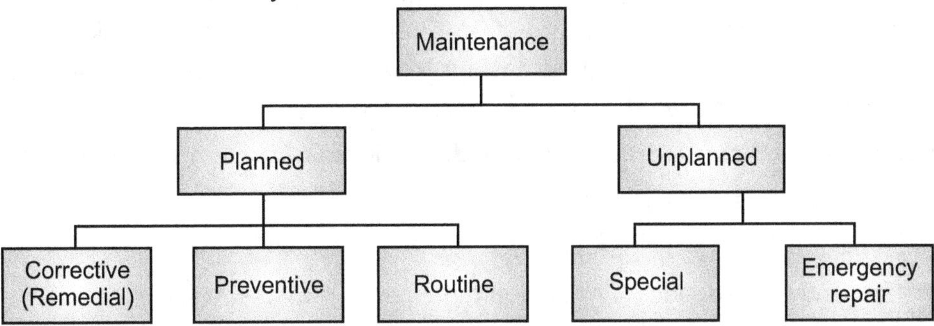

Fig. 1.2 : Classification of maintenance

- **Condition based maintenance :** Work initiated after due inspection.
- **Preventive maintenance :** This is intended to preserve by preventing failure and detecting incipient faults (Work is done before failure takes place)
- **Routine maintenance :** Activities which are repeated at relatively fixed intervals of time.
- **Opportunity or special maintenance :** Work is done as and when possible within the limits of operation demand.
- **Emergency maintenance :** Necessitated by unforeseen breakdown drainage or damage caused by natural calamity like fire, floods, cyclone earthquake etc.

1.2.1 Preventive Maintenance

- Preventive maintenance is defined as, "the maintenance work done before a defect or damage developed in the structure". Preventive maintenance is an active technique to prevent untimely collapse. This includes thorough inspection, planning maintenance programs and executing work. It depends on the specifications, condition and usage of the structure.
- **The purpose of preventive maintenance for civilian structures is :**
  - To protect civil structure from deterioration.
  - To ensure that civil structures adequately support their assigned missions.
  - To prevent the failure of structures before they actually occur.
  - To maintain and enhance the reliability of structures by repairing/replacing damaged structural components before they actually fail.

- There can be no fixed frequency for preventive maintenance of civil engineering structure. Preventive maintenance of civilian structures will be mostly in accordance with the plant requirements and recommendation inspection reports.
- It shall be decided on the following input :
  - Previous structure failure and repair record.
  - Requirement and recommendation of various departments like Mechanical / Electrical / Instrumentation / Plant operation etc.
  - Civil inspection recommendation.
  - External agency recommendation and report based on distress category and NDT results.
  - Design life of structures.
  - Aging mechanism.
  - Consequences of failures.
  - Structural criticality.
- Plant operations as well as civilian inspections will determine whether it will cost more for regularly scheduled downtime and maintenance, than it will normally cost to operate structure until repair is absolutely necessary. This may be true for some structures; while there should be no comparison only on cost, but long term benefits and savings associated with preventive maintenance will also be considered. For example the cost for lost production time from unscheduled structure breakdown will be incurred. Also preventive maintenance will result in savings due to increase of effective system service life. Preventive maintenance also depends on the use of structures.
- **Following points must be considered while deciding course of preventive maintenance :**
  - Function of the structure.
  - Degradation mechanism.
  - Credible effect and consequences (What would happen if the asset fails).
  - Determine condition status (at most recent inspection).
  - Estimate remaining functional Life (as of most recent inspection date).
- Based on above points, preventive maintenance can be taken up keeping reliability as well as type of repair and estimated cost up to the life cycle of structure. As a result, more insight into and control over the maintenance of structures is desired.
- Below three points are important for inspection and maintenance related to structure distress :
  (i) Technology to discover damage in early stage which leads to serious situation.
  (ii) Technology to rightly evaluate present available performance (durability and load resistance performance.
  (iii) Technology to estimate the progress of damage (degradation prediction technique).

1.2.2 Remedial Maintenance

- It is the maintenance done after the defects or damage occurs in the structure. It involves the following basic steps :
  - Finding the deterioration.
  - Determining the causes.
  - Evaluating the strength of the existing structure.
  - Evaluating the need of the structure.
  - Selecting and implementing the repair procedure.

1.2.3 Routine Maintenance

- Routine maintenance is post construction activity, which is required to be attended for the maintenance of the building to prevent its initial decay and to prevent damage to it and prevent it from becoming non-functional.

- It is the service maintenance involved in the structure from time to time. The nature of the work done and the interval of time at which it depends on the specifications and materials of the structure, purpose, intensity and conditions of use.

- This includes white washing, patch repair to plaster, replacement of fittings and fixtures, road surface bonding.

- Regular maintenance of the structure is necessary to keep it functional and to prevent early decay.

- A building is made of different parts in different location and is made of different materials. All of these are susceptible to natural decay due to aging. While designing, the life of the members is assumed with normal maintenance.

- There are various items of work which come under routine maintenance and it is expected that the maintenance of the building will be attended regularly. Some items are required to be attended daily, some weekly, while some at regular intervals.

1.2.4 Special Maintenance

- It is a work done under special conditions and requires heavy clearance and performance to repair heavy damage. This can be done to strengthen and update the structure to meet the new condition of use or to increase its serviceability.

- This may involve particular or complete renovations at long intervals, such as floors, roofs etc.

- Any special repair work to be done in the structure is to be certified by an engineer or licensed structural engineer. The engineer is required to verify himself to the necessity of undertaking special repair to any item of building.

- Special repair estimates will be prepared by the engineer. The governing body or managing committee shall record the estimate in writing.

- There should not be too much estimate for special repairs. As far as possible, the approximate number should be limited to the number of identified subheads.

1.2.5 Extra Ordinary Special Repair

- When expenditure on special repair to a particular building is in excess of the permissible yardstick of special repair, the same come under the category of **extra ordinary special repair**.

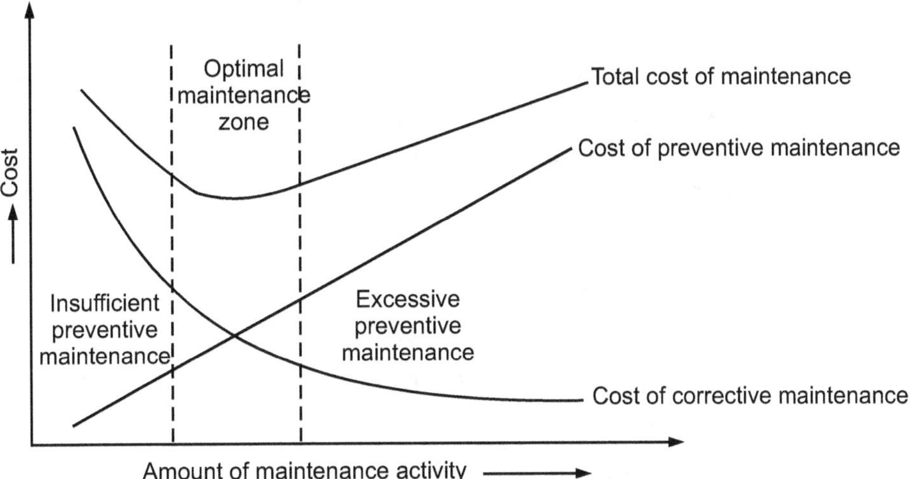

Fig. 1.3 : Regular maintenance to cost graph

- Expenditure on special repair up to permissible limit can be incurred by the Engineer. Beyond the permissible limit however engineer has to have the approval of the higher authorities.

- Governing bodies are empowered to approve the extra ordinary Special Repair Estimate to certain amount so long as scope of expenditure is to retain the building in its original shape in liveable conditions without carrying out any additions to it.

1.3 OBJECTIVES OF MAINTENANCE AND REPAIRS

1.3.1 The Principles of Repair

- **The Purpose of Repair :** The quantum of repair should be kept to the minimum as the main intention is to slow the decaying process of the building. Extra caution must be taken in order not to alter the features of the building. The major aim is to ensure the strength of the structure is able to withstand loads.

- **Avoiding Unnecessary Damage :** The rate in which the decay will take place may vary on the types of material used. Some of the materials, for instant certain type of roof require periodic complete or major replacement. A more selective approach is required at some of these elements such as masonry, framing of walls and roofs. These items will decay slowly and in isolated areas.

- **Analyzing Historic Development :** Owner or contractor involved in preserving the building need to investigate or review the historical data of the building. Usually the main criteria to look into are architectural investigation or any record of particular structure and assessment on its historic context.

- **Analyzing the Causes of Defects :** A deep analysis on the building's historical development, details in design of repairs should be preceded by long term observation. This must include the condition of its material, causes and processes of the rates of decay.

- **Adopting Proven Techniques :** Any repair must be compatible with existing material or methods of construction. This will preserve the integrity to ensure the work done has an appropriate life. If possible, new methods or techniques should not be used unless the old techniques are no longer relevant.

- **Preserving Originality :** All heritage building repairs should be executed to its original nature as possible without trying to hide using artificial ageing. Minimum work is sufficient when heritage building is involved to ensure minimum obstruction to its original built and design.

- **Restoration of Lost Features :** Pinnacles, cornices, hood moulds, window tracery and members of a timber frame that may have been lost in the past. From record we could put these items back in place in the course of repair. Non-structural elements also may be replaced such as railings, windows, rainwater goods or shop fronts. These items may be replaced if we have sufficient data and evidence for an accurate replacement.

- **Removal of Damaging Alterations :** Any renovations, alterations or additions including remedial work done on building are important and to be recorded as a cumulative history of the building. Careful measure and record must be kept and statutory consents must be obtained in advance.

- **Safeguarding the Future :** An interval of 5-7 years should be considered a good, for routine checkup in monitoring building. Any problems can be detected early on such as waterproofing problems which generally started at the highest level. Professionalism is also a good consideration in hiring a third party to maintain the building.

1.3.2 Necessity of Maintenance

- Prevention of damages like decay due to natural agencies, wear and tear.

- To keep structure in good appearance and working condition at all weathering actions.

- To reduce the causes against building.

- To increase the life of building and maintain its value.

- To reduce the risk for occupants and the outgoing expanses.

- To repair the defects occurred in the structure and strength them.
- Adequate maintenance is important one, because repair and re-habitation of structure cost is huge.

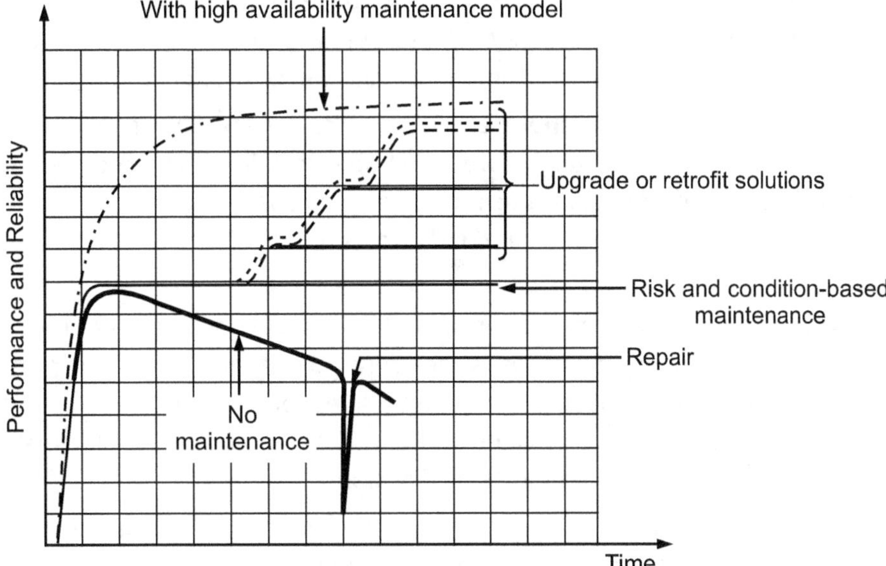

Fig. 1.4 : Structure performance graph

1.3.3 Objectives of Maintenance and Repairs

- The maintenance of structure is done to meet the following objectives :
  - To improve the service life of the structure.
  - For better serviceability of elements and components.
  - This leads to early detection of defects.
  - Prevents major deterioration leading to collapse.
  - Prevention of losses due to natural agencies and keeping them in good appearance and working condition.
  - Repair faults generated in the structure and, if necessary, strengthen them.

1.3.4 Importance of Maintenance and Repairs

- Maintenance management is responsible for the smooth and efficient working of the structure and helps in improving the durability.
- It also helps to keep the structure in their optimum operating conditions. Thus, maintenance is an important and inevitable service function of an efficient structure.
- It also helps to maintain and improve the operational efficiency of the plant facilities and hence contributes to revenue by reducing operating costs and improving the quality and quantity of the product being manufactured.
- As a service function it deals with some cost disturbances. Important components of such costs are - employment of maintenance staff, other small administrative expenses, investment in maintenance equipment and inventory of repair components/parts and maintenance materials.
- Absence of maintenance can lead to repeated breakdowns, collapses and failure of some productive centers/services, resulting in stagnation of activities, idle man and machine time, disorganization of subsequent operations.
- Accidents endanger the life of tenants/public and allied people etc.
- Hence, the absence of planned maintenance service proves costlier. So it should be provided in the light of cost benefit analysis. Since maintenance is a service function, it should be provided at the least possible cost but in a effective way.

1.4 FACTORS INFLUENCING THE MAINTENANCE AND REPAIRS

- Maintenance of the buildings is influenced by the following factors :

    (a) Technical factors : These include the age of the building, the nature of the design, the material specification, the previous standard of maintenance and the cost of postponing the maintenance.

    - **Site selection and site development errors :** Failures often result from unwise land use or site selection decisions. Some sites are more vulnerable to failure. The most obvious examples are sites located in areas of significant seismic activity, in coastal areas, or in floodplains. Other sites face problems related to specific soil conditions such as soil expansion or permafrost in cold regions.

    - **Design errors :** These failures include errors in the concept; lack of structural redundancy; failure to consider load or combination of loads; lack connection; calculation errors; misuse of computer software; including incompatible materials selection, failure to consider maintenance requirements and durability; insufficient or inconsistent specifications for materials or expected quality of works and wage communication of design.

    - **Construction Errors :** Such errors may include excavation and equipment accidents; improper sequencing; insufficient temporary support; excessive construction load; premature removal of shoring or formwork; and non-conformity to design intentions.

    - **Material Deficiencies :** While it is true that most of the material problems are the result of human errors, including a lack of understanding about materials, there are failures that can be attributed to unexpected inconsistencies in the material.

    - **Operational errors :** Failure may occur after occupancy of a facility as a result of owner/operator errors. These may include alteration made in structure, changes in use, careless overloading, and inadequate maintenance.

    (b) Policy : A maintenance policy ensures that, the value of the money spent is obtained in addition to the security of the asset value and the resource value of the buildings concerned and owners.

    (c) Environmental : All buildings are subject to the effects of a variety of external factors such as wind, wind precipitation, temperature etc. which affect the frequency and scope of maintenance. Similar factors of humidity, temperature and pollution will be considered. Industrial buildings can be subject to many different factors. Swimming pool structures are sensitive to the effects of chlorine used in water.

    (d) User : The maintenance requirements of buildings and their various parts are directly related to the type and intensity of their use.

1.4.1 Causes that Affect the Service and Durability of the Structure

1. **Atmospheric Agencies :**

    A. **Rainfall :** It is an important source of water that affects the structure in the following ways.

    - **Physical :** Dissolving and transporting minerals as it is a universal solvent.

    - **Expansion and contraction :** Materials are subjected to repeated expansion and contraction, while they become wet and dry and develop stresses.

    - **Erosion :** Transportation and attrition and abrasion of the materials is quite evident effect of the water.

    - **Chemicals :** Water available in nature contains acids and alkalis and other compound in dissolved form salts on surface of material, known as chemical weathering.

    - **Expansion of water :** The variation of temperature causes expansion and contraction of absorbed water and affects the microstructures of the material.

B. Wind : It is the agent which transports the abrasive material and helps in physical weathering.

- Its action increases, when it is moving with high speed.
- it may contain some acidic gases like CO_2, smoke, which may act over the material and penetrates quite display in the material and structure.

C. Temperature :

- Seasonal and annual variation of temperature.
- Surface of the material causes expansion and contraction.
- The movement of the material bond and adhesion between them occurs when it is repeated.
- The development of cracks and rock may breaker away into smaller units.

2. Normal Wear and Tear :

- During the use of structure it is subjected to abrasion and thereby it losses appearance and serviceability.

3. Causes of failure of structure :

- Improper design due to being incorrect.
- Inadequate data regarding usage.
- Loading and environmental conditions.
- Material selection and poor description.
- Defective construction.
- Poor materials and poor workmanship.
- Lack of quality control and supervision.
- Improper use of structure-over loading.
- Impurities from burning industrial fuels.
- Deteriorating environment.
- Seawater minerals and chemicals.
- Storage of chemicals for which they are not designed.
- Lack of maintenance.
- Lack of proper security, precautions and prevention.

1.4.2 Advantages of Maintenance and Repair

- **Low Risk Factors :** As the building is being regularly checked, the risk of sudden breakdown is low. Hence, creates a safe working environment for residence.

- **Follows a schedule :** By following a schedule, the unforeseen maintenance cost will be minimum and also overall delay in construction activity can be avoided.

- **Long building lifespan :** When the equipment is being tested and maintained, it will be kept in its best shape, so it will extend its lifetime. Along with regular check-ups of building parts such as pipes, boilers and roofing, it will also extend the life of building.

- **Cost effective :** Over a period of time it can be seen that, less money is being spent, because as it will not have to change the equipment much, as well as deal with last-minute break downs. While there still may be some unplanned maintenance required, when the building and equipment are regularly checked. Property-wise, it can be able to detect roof leaks before they escalate and repair quickly before mould and debris occur.

- **Waste less energy :** In general when appliances are not kept in the best conditions possible, they will drain more energy, increasing utility bill. With properly maintained equipment, it will save energy and money. While check-up of the lighting and cooling/heating system on a regular basis will help reduce the energy bill.

- **Less disruption :** With regular checks, it will not be surprised when something goes wrong. This will be a quick fix as user will know what needs to be done. When it comes to closing property and disrupting workers, there will not be problems if a major problem arises.

1.4.4 Limitations of Maintenance and Repairs

- **Require more cost :** When initially starting a preventive maintenance plan, it will cost more to maintain equipment and the building regularly, than to wait for a simply break down.
- **Over maintenance :** Because there is a regular plan, sometimes items often do not need to be checked according to plan. If so, this can change maintenance plan to checking the specific equipment or areas less often, while maintaining a schedule.
- **Require more workers :** Preventive maintenance requires more workers because regular checkups are necessary. When compared with reactive maintenance, users simply need to call expert worker for an onetime fix. Instead this method requires workers to always be on site and perform daily tasks.

1.5 APPROACH TO EFFECTIVE MANAGEMENT FOR MAINTENANCE AND REPAIR

- **Maintenance :** A combination of all technical and associated administrative actions for the purpose of maintaining or restoring an object to the condition in which it can perform its essential functions.
- **Maintenance Management :** Organization of maintenance within an agreed policy. Maintenance can be seen as a 'steady state' activity.
- **Building Maintenance :** The work done to maintain or restore the performance of the building fabric and its services to provide an efficient and acceptable operating environment to its users.
- **Housekeeping :** The regularly recurring work that is required to keep a structure in good condition so that it can be used in its original capacity and efficiency as well as proper preservation of capital investment throughout its economic life.
- **Owner :** Person or body having a legal interest in a building. This includes freeholders, leaseholders or those holding a sub-lease which both bestows a legal right to occupation and gives rise to liabilities in respect of safety or building condition. In case of lease or sub-lease holders, as far as ownership with respect to the structure is concerned, the responsibility of structure of a flat or structure on a plot belongs to the allotee/lessee during the leasehold.
- **Confined Space :** Space which is inadequately ventilated for any reason and may result in a deficiency of oxygen, or a build-up of toxic gases, e.g. closed tanks, sewers, ducts, closed and unventilated rooms, and open topped tanks particularly where gases or vapors heavier than air may be present.

1.5.1 Maintenance Management

- Maintenance management of building is the art of preserving over a long period what has been constructed. Whereas construction stage lasts for a short period, maintenance continues for comparatively very large period during the useful life of building. Though the building shall be designed to be very durable, it needs maintenance to keep it in good condition.
- Inadequate or improper maintenance adversely affects the environment in which people work, thus affecting the overall output and also the overall service life of the building. In the post construction stage, the day to day maintenance or upkeep of the building shall certainly delay the decay of the building structure.

1.5.2 Maintenance Policy

- The policy shall cover such items as the owner's anticipated future requirement for the building, taking account of the building's physical performance and its functional suitability. This shall lead to decisions regarding :
    (a) the present use of the building anticipating any likely upgrading and their effect on the life cycles of existing components or engineering services; and
    (b) A change of use for the building and the effect of any conversion work on the life cycles of existing components or engineering services.

1.5.3 Maintenance Work Program

- Maintenance work will be done at a time when any adverse effect on output or function will be likely to be minimized as well as reasonable consideration for the comfort of residents and public and third party stakeholders.

- The program will be planned to remove any unsuccessful work. This may occur if upgrading or conversion work is done after the maintenance work has been completed.

- Any delay in fixing the defect will be kept to a minimum, if such delay is likely to affect the output or function. Maintenance costs increase with decreasing response time.

- Maintenance work, completed or being performed shall comply with all statutory and other legal requirements.

1.5.4 Maintenance Guides

- An owner responsible for a large number of buildings has to establish procedures for maintenance. When an owner is responsible for the maintenance of only one building or a small number of buildings, drafting a guideline book tailored to suit each particular building can provide significant advantages. Such a manual would take into account the following :

  (a) Type of construction and the type of residual life of a building, and

  (b) Environment and intensity of use.

1.5.5 Planning of Maintenance Work

- Work shall take into account the possible maintenance cycle of each building element and logically planned with inspections being done at regular intervals. Annual plans shall take into account subsequent years' programmed to incorporate items and to prevent additional cost. It shall be emphasized that, the design of some buildings can lead to high indirect costs in maintenance contracts and, therefore, careful planning can bring financial benefits. The decision to repair or replace shall be taken after due consideration.

- Identifying the exact need of the customer/owner, that is, having a clear idea about the various items of work that would be required for efficient rehabilitation of the structure as desired along with their sequence.

- When above stages are complete, that is, the items of the work are identified, they have to be surveyed on site by an experienced and skilled Quantity Surveyor to prepare cost estimates along with economic analysis.

- The estimate thus prepared helps in preparing various schedules and finalizes the work plan in consultation with the client about the probable date of commencement of work, time period which will be available for completion and status of the fund.

- Depending on the planning, schedules will be prepared for various items. Schedules of items and quantities of work along with their sequence is very important in planning. A complete schedule of the various materials that will be required and the steps of the requirement match the sequence of activities. Schedule of plants and machinery as would be required to match with the schedule and sequence of items of works. There are other schedules also e.g. labour schedule, expenditure schedule etc.

- Planning and scheduling are activities before the start of the actual work, while control is the activity which occurs during the progress of the work. The purpose of control is to monitor the progress of the work and to see if the work is progressing at all stages desired and planned.

- There are various methods or techniques to control a work. Commonly adopted techniques are bar charts and project evaluation and review techniques, commonly referred to as PERT with Critical Path Method or CPM.

- **Bar Chart :** A bar chart is a chart which shows the progress of various activities in relation to time. The project is divided into several explicit activities, which can be identified separately. These activities are presented linearly against the vertical axis of a two co-ordinate axes diagram. The horizontal axis represents the time elapsed. Each bar represents one specific job or activity of project. The beginning and end represent time of commencement and the time of finishing of the activity.

- The length of the bar, therefore, represents the expected duration of the activity. A bar chart is represented in two axes, time against activity, probable expenditure and actual financial progress also can be shown on the chart along with the horizontal axis. The line on the chart generally takes S-shape and is called S-curve. Thus, a bar chart can represent both physical and financial progress against the program.

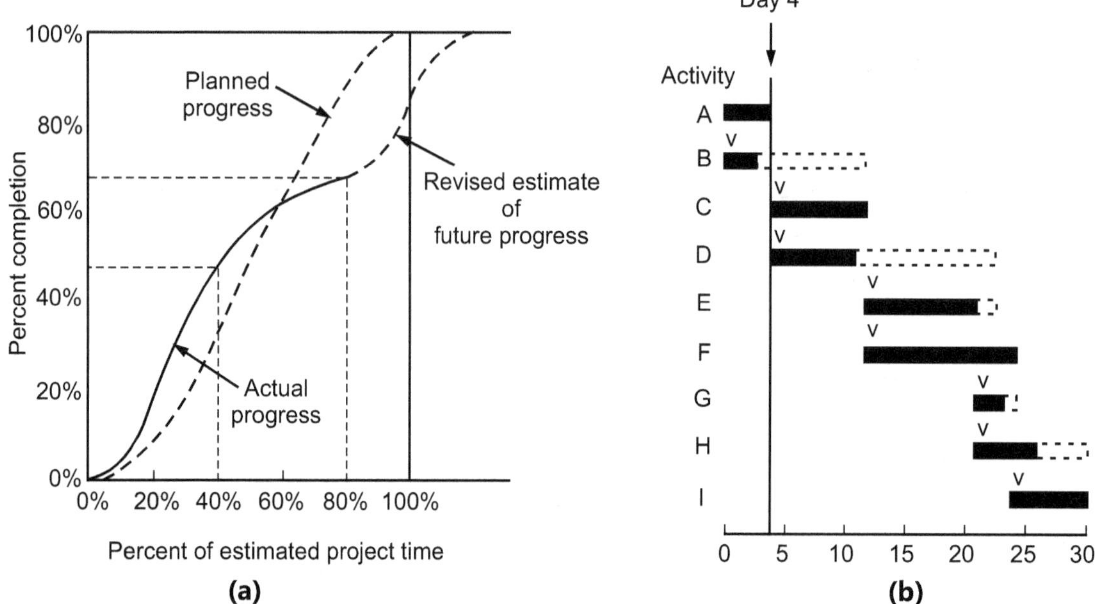

Fig. 1.5 : Physical and financial progress

1.5.6 Service Centre

- A network of service centres is made available for receipt and disposal of maintenance complaints which are made by the occupants. Usually, maintenance engineer is in-charge of the service centre. A drill of maintenance activities to be performed daily, weekly, monthly, annually and periodically by the functionaries.

- The maintenance and repair committee is responsible for planning and supervising all maintenance and repair activity in the building. The committee must work closely with the tenants in order to figure out how much money is available for repairs, and how it should be used.

- Members of the maintenance and repair committee must be building-residents who belong to the tenant association, and they should be prepared to invest considerable time and energy in this important responsibility. The committee should meet at least twice a month to review building repair needs and set priorities.

- When an outside contractor is required, the committee seeks qualified contractors, solicits and evaluates bids for the job, and recommends the selection of a contractor to the executive committee. Members of the maintenance and repair committee usually supervises the work of building's Engineer.

1.5.7 Complaint Register

- Complaint register is an important document maintained at Service Centres. All complaints received at Service Centres are entered in the Complaint Register and these are closely watched to ensure that the complaints are attended to as expeditiously as possible. A typical page of the Complaint Register is given in Table below. There may be different registers for different disciplines for the convenience of concerned Engineer. Complaints can be written by tenants at Service Centers in the form prescribed.

COMPLAINT REGISTER

M/C Number

Sr. No.	Time of complaint	Colony, Flat/ House no.	Description of complaint/ Observation by JE, AE or inspecting officer	Classification (tick where appropriate)		Action Taken				Remarks of I.O. balance work to be done	Balance work carried to complaint no. .../ Transferred to	Sign of I.O.	Remarks of engineer
				No delay	Other	Given to	Date given	Date attended	Details of work done				

- Every complaint shall be assigned a serial number. Time of lodging of complaint shall be invariably recorded by the Receptionist. The complainant shall be intimated with the complaint number and the likely time frame for attending the complaint for his reference. All complaints shall be entered in the register. Civil, Plumbing, Electrical and Horticulture complaints may be entered in different registers.

- First entry in the complaint register on any day shall start on a new page. In case a complainant desires to lodge a written complaint, he shall be given acknowledgement of the complaint in the counterfoil of the slip. As far as possible, the tenants shall be asked to indicate the nature of complaints to the attendant at Service Centre, so that the right person is deputed for the job and he carries with him necessary tools and materials.

- In many cases, the complaints are vague and the workmen have to make more than one trip to the house, to find out the nature of work and the tools and materials required. This wastage of man power should be avoided.

- Tenants shall be advised to register their complaints invariably with the Service Centre. The tenants may be advised to approach the maintenance committee or governing cell only in cases where the complaints are not attended to within a reasonable time or the work has not been done satisfactorily. They shall be asked to quote the complaint number and date, so that the complaint could be investigated.

- Maintenance information system envisages submission of maintenance complaint returns by the Engineers. It will include under mentioned three categories of complaints.

1. Day to day Complaints :

- Engineers will send the day to day complaints to the maintenance committee respectively, separately for all the service centres under them. The day to day complaints will be grouped trade worker wise as under :

Civil	Electrical
1. Plumber	1. Electrician
2. Sewer man	2. Lift
3. Mason	3. Pump/Motor
4. Carpenter	4. AC/Coolers
5. Others	5. Fire alarm
	6. Wet riser
	7. Miscellaneous

- These complaint groups will be prepared by the governing body separately in the complaint registers. Total of all the complaints however will be indicated in the return. The remarks column will indicate the type of pending oldest complaint along with the reasons for the delay and assistance required. Engineers will indicate these complaints in the return formats separately.

2. Special Repairs and Additions/Alterations Complaints :

(a) Special Repairs Complaints :

- Total number of complaints received will include the complaints received from the tenants, as well as observed by the technicians/engineers/officers during their inspections and surveys conducted. The complaints will be indicated group wise. Groups of Civil and Electrical complaints are indicated as under :

Civil	Electrical
1. Structural repairs i.e. concrete work, brick work etc.	1. Wiring/switches/light/power points
2. Finishing woks i.e. plastering, flooring etc.	2. Fittings and fixtures
3. Wood work	3. D.B., panels, controls
4. Steel work	4. Window coolers/AC
5. Sanitary and water supply	5. Lift/pump/generator
6. Water proofing treatment	6. AC plant package plant

DAY TO DAY COMPLAINTS

(To be reported by Engineer Monthly)

Sub-division :

Division   :   **Month :**

Sr. No.	Service centre	No. of complaints received		No. of complaints attended in the month	Complaints pending at the end of the month		Remarks - Reasons for delay in attending the complaints, Assistance required
		B/F from previous month	Received during the month		No.	Date of oldest complaint	
1	2	3	4	5	6	7	8

SPECIAL REPAIRS AND ADDITIONS/ALTERNATIONS COMPLAINTS
(To be reported by Engineer Monthly)

Sub-division :

Division   :   **Month :**

No.	Group of repairs/ additions and alterations	No. of complaints received		Complaints attended (No.)			Complaints pending		Remarks - Reasons of delay, Assistance required
		B/F from previous month	Rcd. in the month	Target for qtr.	Attended in the month	Attended in the qtr.	No.	Date of oldest complaint	
1.	2.	3.	4.	5.	6.	7.	8.	9.	10.
	X--Service Centre (a) Special Repairs (i) (ii) (iii) (b) Additions/Alterations (i) (ii) (iii) **Y-Service Centre** (a) Special Repairs (i) (ii) (iii) (b) Additions/Alterations (i) (ii) (iii)								

- Remarks column will indicate the group of oldest pending complaint, reasons for delay, assistance required.

(b) Additions/Alterations Complaints :

- Similar system as for special repairs complaints will be followed for additions/alterations complaints also.

1.5.9 Modalities of Maintenance

- Whether the work should be carried out through contract or own work force is decided on the nature of the following :
  - (a) Type of work.
  - (b) Quantity of work.
  - (c) Expediency or urgency of work.

Maintenance works are done through one of the following :

1.5.9.1 Through Directly Employed Labor

- Directly employed labor is best suited for day to day routine maintenance. The cost of installation is directly charged to the maintenance work due to their salaries, allowances etc. to the workers. The day to day work charged employees are only for maintenance of buildings. Work charged employees are generally to be employed for operations and routine maintenance work.

- Maintenance and repair work of a special and complex nature that is not usually covered by the duties of the working staff may be done through contractors. Annual repairs such as color washing, distempering, painting, whitewashing etc. additions and alterations to works or minor works that do not require immediate execution, are usually not covered by their duties and can be done through contractors.

- Casual labor is employed to do departmental work. The muster roll is issued for a specified period of time, which usually does not exceed one month at a time and for a specific work. Measurements are recorded if the work is susceptible to measurement. Wherever there is a slight decrease in the strength of the work charged establishment, it is made good by appointing muster roll personnel.

1.5.9.2 Through Contract

- Annual repair works such as white washing, painting, petty work such as replacing glass panes, repairing plaster, replacing roof tiles etc. are usually done through contract. Special repair works such as water proofing treatment work, repair of water supply pump sets, equipment and accessories of AC plants, substation equipment, DG sets, lifts, fire alarm detection and firefighting systems etc., are of good magnitude in financial terms, so these are usually done through contracts. It would be better to appoint special contractors for carrying out the works.

1.6 PERIODICAL MAINTENANCE

- As the building ages, various parts of the building and services decline. Major repair and replacement of elements becomes unavoidable. It becomes necessary to restore the structure until it becomes necessary to prevent the structure from deterioration and improper wear and tear as well as to restore it back to its original condition to the extent possible.

  The maintenance of a general term can be identified into the following broad categories :

  - **Cleaning and servicing :** This is largely a preventive type, such as checking the efficiency of rainwater gutters and servicing mechanical and electrical installations. It also covers the house keeping.

  - **Improvements and repairs :** This includes periodic maintenance work undertaken by, annual contracts and including exterior re-plastering, internal finishing etc.

  - **Replacement :** This includes major repairs or restorations such as reproofing or re-construction of faulty building parts.

1.6.1 Item of Work Which Need to be Attended Weekly

(i) The roof top should be cleaned weekly as otherwise dust and rubbish would block the outlets causing accumulation of rain water on the roof, which ultimately would find way through the roof causing severe structural damage. Sweep off any standing water after rains.

(ii) Bathrooms and bathing places should be cleaned by flushing with 2/3 buckets of hot water at least once in a week. This will loosen up oil and fat particles clogging the trap. Earth and ashes should not be used for cleaning the utensils as this would cause chokage of the trap and, ultimately, shorten its life.

(iii) The doors and windows may give uneasy sound of hinges indicating oiling is required. The hinges should be oiled once in a week. Glass panes of doors and windows are to be cleaned properly with the help of liquid cleaners available commercially.

(iv) The ventilation installations need to be checked, cleaned and oiled once in a week.

(v) The decorations inside and outside are to be cleaned properly at least once in a week.

1.6.2 Items of Work Which Need to be Attended Periodically

(i) Leaks may be observed in soil pipes, waste water pipes and rain water pipes, especially in portions running horizontally. This may be due to leakage through the joints. The portion of the pipe leaking should be closed, and damaged joint opened and cleaned. The joint then shall be redone as before. In case of CI pipes, the joint should be by lead caulking and in case of asbestos pipes, the joint shall be by cement.

(ii)   The water supply lines are to be checked and, in case of any leakage observed in the line, the portion is to be taken out and cleaned by dilute commercial acid and brush and finally washed by clean water and refixed.

(iii)   Cleaning of the water tanks, both at ground and overhead, is essential for hygienic reasons. These should be periodically cleaned at intervals of three to six months.

(iv)   Narrow hair cracks may be observed in the walls. These should be dug up and filled with cement mortar/crack sealer. This would prevent further damage of the affected portion. These filled-up cracks shall be kept under observation.

(v)   If growth of small plants may be observed on the wall, these shall be stopped when they are small. Otherwise, they would create cracks in the adjoining portion of the wall and would lead to major trouble in future. The plants shall be uprooted and the place should be treated with copper sulphate solution or acid for permanent eradication.

(vi)   Plastering on the walls both internal and external and ceiling may show bulging or cracks. These areas must be checked thoroughly by tapping with a light wooden/PVC hammer. The portions emitting dull sound indicate separation of the plaster from the surface. These portions shall be taken out in regular shape and replastered with mortar of same proportion after raking out the brick joints and cleaning the surface.

(vii)   Painting internal and external surfaces of building is essential for various hygienic reasons and protection of structure and aesthetic. The rendering on the surface protects the structure. But it is porous and absorbs moisture which causes permanent damage in the walls and ultimately affects the structure seriously. The external painting seals the pores of the plaster and protects the structure. Renewal of the internal wall painting with cement paint is desired annually and that of the external walls every fourth year. Use of Acrylic paints will decrease the frequency of painting.

(viii)   Painting of doors and windows and water supply and drainage lines is to be done periodically at intervals of not more than four years.

(ix)   The insulations either sound or heat, if exist, are to be checked against any leakage. If observed, should be repaired immediately.

(x)   Electrical installations, internal wiring, switches, fans, water-heater etc. are to be checked to find out if there is any leakage spot which is common in old buildings. These are to be cleaned at regular interval.

(xi)   Plinth protection around the building need to be maintained properly so that there may not be any passage for the surface water to percolate to the foundation threatening its settlement.

(xii)   There may be installations like lifts, escalators, service elevators for vertical transportation of persons or goods. After installation of these, Servicing and Maintenance Contracts are entered with the firms installing these. They do the maintenance job at regular intervals.

(xiii)   Cleaning of the premises, i.e., compound of the building including cutting and removal of unwanted plants, shrubs etc. and removal of garbage, is to be done at regular intervals to keep the area clean and pleasant.

Roof	• Replace or repair loose or missing coping stones. • Check for cracks or blisters in the roofing, and repair as needed.
Exterior Walls	• Maintain gutters and down spouts to keep water off exterior walls. • Check condition of mortar between bricks, and repair as needed. • Check condition of weather seal around windows, doors and skylights, and repair as needed.
Windows and Doors	• Keep all wooden parts freshly painted. • Check putty and replace as needed.
Plumbing	• Repair leaky joints as they are reported. • Check for plumbing leaks and repair immediately. • Clean out sink traps and building trap (in basement) if draining is slow.

1.6.3 Special Maintenance

- White washing, colour washing, distempering etc., after completely scrapping the existing finish and preparing the surface.
- Painting after removing existing old paint from various members.
- Provision of water proofing treatment on the roof. All existing treatments known are supposed to last satisfactorily for a period of about ten years.
- Repair of internal roads and pavements.
- Repair/replacement of flooring, skirting, dado and plaster.
- Replacement of doors, window frames and shutters. Replacement of door and window fittings.
- Replacement of water supply and sanitary installation such as water tanks, WC cisterns, wash basins, kitchen sinks, pipes etc.
- The expected economic life of the building under normal occupancy and maintenance conditions is considered as follows :
  - Monumental buildings 100 years.
  - RCC framed construction 75 years
  - Load bearing construction 55 years.
  - Semi-permanent structures 30 years.
  - Purely temporary structures 5 years.
- The life of the above mentioned buildings is only indicative and depends on many factors such as location, use, specifications and maintenance.
- Building services, fixtures including internal wiring, water supply distribution system etc. are expected to last for 20-25 years. Later, it may be necessary to replace them after detailed inspection. Electrical repairs in general are whole sale replacements of the wiring electrical installations. Earthing also has to be attended.

MAINTENANCE NORMS, FREQUENCY OF APPLICATION OF FINISHING ITEMS

Sr. No.	Item	Periodicity				
		Res. Bldg.	Office Bldg.	Hospitals	Laboratories	Schools
1	2	3	4	5	6	7
1.	White washing/Colour washing	2 years	2 years	2 years	2 years	2 years
2.	Applying dry distemper	2 years	2 years	2 years	2 years	2 years
3.	Painting with plaster paint, Synthetic enamel paint, Oil bound distemper, Acrylic paint, Acrylic distemper	3 years	2 years	1 year - Corridor O.T. Rooms 2 years - Other areas	2 years	3 years
4.	Painting external surface with water proofing cement paint	3 years	3 years	3 years	3 years	3 years
5.	Cleaning and disinfecting of water storage/ distribution tanks, water mains.	6 months	6 months	3 months	3 months	6 months
6.	Cleaning of manholes/gully chambers/inspection chambers and flushing of building sewers	1 year	1 year	6 months	1 year	1 year
7.	Cleaning of storm water drains	1 year	1 year	1 year	1 year	1 year

… Contd.

8.	Painting steel water tanks inside with bitumastic paint	2 years	2 years	1 year	2 years	2 years
9.	Polishing wooden doors/windows with sprit polish/ polish/synthetic acrylic polish	5 years	5 years	5 years	5 years	5 years
10.	Text mat or poly mat based equivalent synthetic silicon based exterior paint	5 years	5 years	5 years	5 years	5 years
11.	Cleaning electrical installations, fans etc.	1 year	1 year	1 year	1 year	1 year
12.	Premix, semi dense/dense carpeting of roads	5 years	5 years	5 years	5 years	5 years
13.	Collection of water samples for physical, chemical and bacteriological analysis of water	6 months	6 months	3 months	6 months	6 months

1.6.4 Corrective Maintenance Procedures and Sources

- Upon discovering an additional problem, corrective maintenance is planned and scheduled for future times. During the execution of corrective maintenance work, the item is repaired, restored or replaced. Within the maintenance area, corrective maintenance begins when a technician sees something that is about to break or affect the overall performance of the structure. It can still be repaired or reinstalled without downtime.

- If corrective maintenance is not scheduled, the problem can become an emergency maintenance work down the road and result in evacuation of tenants, service interruption or unhappy customers. For example, if technicians are performing preventive maintenance for plumbing lines and notice significant leakage or damage on a critical part or component, corrective maintenance can be initiated or that part can be restored as soon as possible.

- In a public work situation, it can make regular roadway repairs while noticing some signage damage from a recent storm. This can later initiate a corrective maintenance order to restore that signage.

Advantages of Corrective Maintenance

Since corrective maintenance is performed "just in time", the main benefit is a reduction in emergency maintenance orders as well as employee safety.

- Corrective maintenance work orders are set and prioritized, which helps maintenance teams solve problems before a disaster.

- Corrective maintenance, combined with good preventive maintenance, helps a business to extend the lifetime of its assets, reduce employee injury and optimize resource planning.

- Corrective maintenance work orders are often less expensive to implement than requiring emergency maintenance work orders to be completed during overtime hours.

How to Maximize Corrective Maintenance ?

- Corrective maintenance has an important role within the scope of maintenance services in a business, residential complex, or factory setting. A qualified and experienced technician can find and identify potential problems before they become an emergency.

- To maximize the benefits of corrective maintenance, provide training and education to maintenance technicians and supervisors on what types of things to look for while they are providing preventive or emergency maintenance services.

- Additionally, create a maintenance checklist of components that technicians can visually check quickly. This will generate more corrective maintenance orders and help the facility run smoothly.

1.6.5 Maintenance Manual (Register of Buildings)

- Each building/society/institution will maintain a register of buildings till date. The appointed maintenance engineer will certify that effect at the end of every financial year after ensuring the necessary additions to the cost and structures. This certificate will be presented by them to the governing body every year. Garden registers may be maintained by a horticultural expert, indicating in original works, additions and alteration works and special repairs, if necessary.

- The following building records will be maintained by the governing body in the manual :

Name of Building/Institute /Society
Location & Registration no :
Salient details of the building
1. W/S and Sanitary installations
(a) Water supply :
(i) Sources of W/S to the building and source wise capacity.
(ii) Brief specification of W/S distribution system.
(b) Details of tube wells if any :
(i) No. of tube wells.
(ii) Yield of each tube well.
(iii) Type of boring and depth.
(iv) Brief specifications of boring pipes.
(c) Type of w/s distribution :
(i) OH tank details like material, lining, height, capacity etc.
(ii) W/S sump details like material, depth, capacity etc.
(d) Sewerage System :
(i) Specification of sewerage system.
(ii) Out fall of sewerage system.
(iii) Sewage sump, details if any.
(iv) Sewage treatment plant details if any.
2. Electrical and Air Conditioning System
(i) Details of internal electrical installations like wiring etc.
(ii) H.T. and L.T. Pannels details
(iii) Type of fans
(iv) Whether the building is centrally air conditioned.
(v) A.C. load
(vi) Type of plant
(vii) Brief specifications
3. Firefighting/Fire alarm system
(i) Type of firefighting system.
(ii) Type of fire alarm.
(iii) Location of control room.
(iv) Brief specification of firefighting system.
(v) Brief specification of fire alarm system.
(vi) Fire tank capacity :
(a) Over head tank.
(b) Under ground tank.
(vii) No. of fire hydrants.
(viii) Fire pumps : No. and capacity.

4. Details of Guarantee/Warranty certificates obtained for building for various services/treatments
5. Contract details
(a) Name of the building contractor and address
(b) Agreement
(c) Completion period
(i) Commencement date
(ii) Date of completion
(d) Approximate completion cost :
(i) Building
(ii) W/S and Sanitary and fire fighting
(iii) Electrical and Air conditioning
6. Major defects noticed at the time of handing over/taking over
(a) Building work.
(b) W/S and Sanitary installations.
(c) Electrical and AC works.
7. Handing over of completion drawings (4 sets of each)
(a) Architectural drawings numbers.
(b) Structural drawings numbers.
(c) Water supply, sewerage and drainage drawings numbers.
(d) Electrical & AC drawings numbers.
8. The manual in service centre shall also have
(a) Contact details of the professionals involved during the construction of building.
(b) Contact details of the professionals/technicians for preventive and emergency maintenance.
(c) The sources and test reports of the material for emergency.
(d) Well-equipped safety team.

1.6.6 History Sheet of Maintenance

Repair estimate :

- Annual repairs and maintenance estimates for buildings and Services are prepared in the beginning of the year. The estimates cater to day to day repairs and annual (periodical) repairs and should include the whole expenditure on cost of labour, cost of materials required for day to day works, cost of work being carried out through work orders and contracts, anticipated to be incurred during the working year on the maintenance of buildings in question. The total estimated cost of maintenance of buildings/structures during the year should be within the prescribed limits as approved by the Governing body concerned from time to time both for annual repairs and special repairs.

Register of periodical repairs :

- This register shall be maintained in form given below. Complaints of periodical nature like white washing, painting etc. which are usually got attended through contractors and cannot be attended to on daily basis are transferred to this register. From this register/records of the particular premises appropriate information shall be passed on to the complainant about the admissibility of the request and the likely time it shall require for the compliance.

REGISTER OF PERIODICAL REPAIRS

Sr. No.	Complaint No.	House No./ Locality	Request regarding		Due/Not due	Date planned for the work	Date of intimation to allottee	Date of completion of work
			White washing/ Distempering, Fans cleaning, Surface dressing of lawn, Vegetation cleaning etc.	Door/Window paining, Painting of fans, D.B. open metal conduits etc.				
1	2	3	4	5	6	7	8	9

Register of special repairs :

- This register shall be maintained in form given below. Complaints of special nature repairs, which cannot be attended on daily basis, shall be transferred to this register. The special repairs to buildings may be divided in following groups :
    - Concrete work.
    - Masonry works including plaster, flooring and brick work.
    - Wood work.
    - Steel work.
    - Sanitary and water supply.
    - Water proofing treatment.
    - Electrical wiring and fittings.
- Few pages shall be allotted separately to each of these groups in the register and an index shall be prepared in the beginning of the register. The complaint of special repair nature shall be transferred from the complaint register to the relevant group in this register. All details about the complaint shall be properly filled in the columns of the register.

REGISTER OF SPECIAL REPAIRS

Special Repair Group ....

Sr. No.	Complaint no.	House no./ Locality	Location of repairs	Approximate quantity	Repairs required in			Schedule of repairs
					Less than 1 month	Within 3 months	Within 6 months	
1	2	3	4	5	6	7	8	9

1.6.7 Pre and Post Monsoon Maintenance

Pre-monsoon Period :

- Pre monsoon inspection are made do decide the maintenance program to be done before monsoon such us cleaning of drains, checking of root leakage, collection of material and equipment required during monsoon repairs.

- Before the onset of the monsoon, especially those structures in the area of heavy rainfall should be inspected for taking advance action as a preventive measure to stop or at least reduce the damaging effect of monsoon.

- Probable items of inspections may be.

  o The roofs shall be inspected thoroughly. If any doubtful spot is detected, the same shall be repaired. If no definite spots could be identified, but the surface appeared to be vulnerable, a coat of bitumen or waterproofing chemical would be of help.

  o If active, dormant or dead cracks are observed, preventive measures in sealing the cracks shall be taken. Any damage found and separation of the brick paving from the plinth shall be repaired.

  o Vegetation needs to be removed, the roots extracted and the place shall be treated with chemicals for stopping further growth.

  o In pitched roof the covering sheets, joints, overlaps, fixing screws shall be checked and repaired, if found vulnerable. The roof drainage system and drainage pipes shall be checked and cleaned wherever found necessary.

Post monsoon inspection :

- Immediately after the monsoon, all structures exposed to weather shall be checked to find if there had been any damage due to the monsoon. Preliminary inspection shall be undertaken by a group of technical staff to assess in a general way.

- The various damages and decays the building has suffered are identified without going into the details and intricacies of the damages. Assessment of damages are to be done by visual inspection. There may be concealed portions of the structural members which may not be accounted for in the inspection.

  (a) Sanitary and water supply installations water closets, washing places, baths etc.

  (b) Common passages, verandahas, balconies.

  (c) Wall painting, both internal and external.

  (d) Inadequate drainage and accumulation of water on roof.

  (e) Cracked walls.

  (f) Settlement of foundation.

  (g) Decay and damages.

  (h) Warping of timber works.

  (i) Any structural crack indicating structural disbalance etc.

Important Points

Repair	Without strength.
Retrofitting	Increasing strength to carry more than designed strength.
Re-strengthening	Re gaining strength.
Rehabilitation	Repairing the structure for serviceable condition that is current loads.
Restoration	Restoring the original strength of structure for designed loads.

- Maintenance is preventive in nature. Activities include inspection and works, necessary to fulfill the intended function or to sustain original standard of service.

- Repair is the technical aspects of rehabilitation. It refers to the modification of a structure, partly or wholly, which is damaged in appearance or serviceability.

- Rehabilitation is the process of restoring the structure to service level, once it had and now lost, strengthening consists in endowing the structure with a service level, higher than that initially planned by modifying the structure not necessarily damaged structure.

- Corrective maintenance are maintenance tasks that are performed in order to rectify and repair faulty procedures and components. The purpose of corrective maintenance is to restore damaged objects. Corrective maintenance is initiated when an additional problem is discovered.

- Preventive maintenance is a proactive technique to prevent untimely breakdown. Preventive maintenance is planned maintenance actions aimed at the prevention of breakdown or failures.

- The maintenance of structure is done to improve the service life of structure, for better serviceability of elements and components. It leads to quicker detection of defects. It prevents major deterioration leading to collapse. It prevents damages due to natural agencies and keep them in good appearance and working condition. It is done to repair the defects occurred in the structure and strengthen them, if necessary.

- Maintenance management of building is the art of preserving over a long period what has been constructed. Whereas construction stage lasts for a short period, maintenance continues for comparatively very large period during the useful life of building. Inadequate or improper maintenance adversely affects the environment, in which people work, thus affecting the overall output and also the overall service life of the building.

Questions for Practice

1. Define Maintenance.
2. Define Repair.
3. Define Rehabilitation.
4. What are the facts of maintenance?
5. Write down the importance of maintenance.
6. What are the various types of maintenance ?

❑❑❑

Chapter 2

CAUSES AND DETECTION OF DAMAGES

Weightage of Marks = 10, Teaching Hours = 10

Syllabus

2.1 Causes of damages due to distress, earthquake, wind, flood, dampness, corrosion, fire, dilapidation, terminates.

2.2 Systematic approach of damages detection, various aspects of visual observations for detection of damages.

2.3 Tests on damaged structures : rebound hammer, ultrasonic pulse velocity, rebar locator, cover gauge, crack detection microscope, chloride test, sulphate attack, pH measurement, half-cell potential meter.

Objectives

After learning this chapter, student will be able to

- Explain the causes of damages occurred in the given structures.
- Know the systematic approach of damages detection.
- Explain non-destructive test with damaged structural elements.
- Explain the procedure or non-destructive test used to detect the damages.

INTRODUCTION

- During last few decades many structures have suffered significant deteriorations, sometimes severe ones, after a relatively short service time, as consequence of the action of aggressive environments.

- The practice of a systematic investigation of the structure placed in aggressive environments permits both the detection of dangerous situations of damage, and the decision to undertake immediate measures of intervention with the aim to prevent causalities, important material losses, technological accidents or undesired ecological effects, as well as to stop accelerated deterioration processes on the members/structures.

- When making a diagnosis concerning intervention measures for the rehabilitation of the structures, one must consider not only the effect on the building materials (concrete, reinforcement), but also the broader issues of design and execution of intervention works for different categories of structures.

2.1 NATURAL CAUSES OF DAMAGES OF A STRUCTURE

2.2.1 Causes of Deterioration of Concrete Structures

(a) Design and Construction flaws :

- The design of concrete structures controls the performance of concrete structures. Well-designed and detailed concrete structures will show less deterioration compared with poorly designed and detailed concrete, in similar condition. Beam-column joints are particularly at risk of defective concrete, if detailing and placing reinforcement is not carried out properly. Inadequate concrete cover may lead to carbonation depth reaching up to reinforcement, increasing the risk of reinforcement erosion.

- The construction work should be done as per the prescribed specification. The strength, permeability durability of concrete is controlled by adhering to the specified water-cement ratio. Inadequate vibration can result in porous and honeycomb joint concrete, while excess vibration may lead to segregation.

(2.1)

(b) Poor quality material used :

- Quality of materials to be used in construction, should be ensured by means of various tests as specified in the IS codes. Alkali-aggregate reaction and sulphate attack results in early deterioration. Clayey materials in the fine aggregates weaken the mortar aggregate bound and reduce the strength. Salinity causes corrosion of reinforcing bars as well as deterioration of concrete.

(c) Accidental loading :

- Accidental loading can be characterized as the effect of short duration, one-time events such as impact of barge against a lock wall or an earthquakes. These loadings can generate stresses greater than the strength of the concrete, resulting in localized or general failure.

- Determining whether accidental loading caused damage to concrete would require knowledge of the preceding events of damage. Typically, it would be easy to diagnose the damage caused by accidental loading.

(d) Environmental effect :

- Micro cracks present in concrete are sources of moisture. Atmospheric carbon di-oxide penetration into the concrete attacks reinforcement with various ingredients of concrete. Aggressive environment in concrete structure will be reduced severely.

(e) Corrosion of embedded metals :

- Steel reinforcement is intentionally and almost always placed within a few inches of a concrete surface. In most circumstances, portland cement provides good protection to concrete embedded reinforcing steel. This protection is usually attributed to the high alkalinity of the concrete adjacent to the steel and the relatively high electrical resistance of the concrete. Nevertheless, corrosion of reinforcing steel is one of the most frequent causes of damage to concrete.

(f) Abrasion :

- Abrasion damage is caused by waterborne debris and the techniques used to repair the damage on several corps' structures. Abrasion-corrosion damage is caused by rolling debris and grinding against a concrete surface. In hydraulic structures, the areas most likely to be damaged are spillway aprons, stilling basin slabs and locked culverts and lateral.

- Sources of debris include construction waste left in a structure, brought back to a basin by eddy currents due to poor hydraulic design or asymmetric discharge, and riprap or other debris dumped into the basin by the public. Also, friction erosion can occur from impacting or scraping barges and towels on pond wells and guide wells.

(g) Cavitation :

- Cavitation-erosion is the result of relatively complex flow characteristics of water on concrete surfaces. Cavitation damage can occur when the flow is quite fast (more than 12.2 m/s) and where there is surface irregularity in the concrete.

- Whenever there is an irregularity in the surface, the flowing water will separate from the concrete surface. Vapor bubbles will develop in the area of separation from the concrete as the vapor pressure in the area will decrease. As these bubbles are moved downstream, they will soon reach areas of normal pressure. These bubbles will collapse with an almost instantaneous decrease in volume. This collapse, or entrapment, creates a shock wave which, upon reaching a concrete surface, generates too much stress over a small area. Repeated falls of vapor bubbles on or around the surface of the concrete will cause exhaustion. The concrete spillways and outlet works of many high dams have been severely damaged by cavitation.

(h) Freezing and thawing :

- As the temperature of critically saturated concrete decreases during the cold weather, the freezing water held in capillary pores of the cement paste and aggregate expand upon freezing. If subsequently thawing is followed by refreezing, the concrete is further expanded, so that the cycle of repeated freezing and thawing has a cumulative effect.

- By their nature, concrete hydraulic structures are particularly vulnerable to freezing and thawing because there is ample opportunity for parts of these structures to become critically saturated. Concrete is particularly vulnerable to water level fluctuations or spraying conditions. Exposure in such areas as the top of walls, piers, parapets, and slabs increases the vulnerability of concrete to the damaging effects of repeated cycles of freezing and thawing. The use of chemicals on concrete surfaces can accelerate the damage caused by freezing and thawing and can lead to exhaustion and scaling. This involves the development of osmotic and hydraulic pressure during freezing, principally in the paste, which is similar to ordinary frost action.

(i) Settlement and movement :

- Situations in which the various elements of a structure are moving with respect to one another are caused by differential movements. Since concrete structures are typically very rigid, they can tolerate very little differential movement. As the differential movement increases, concrete members can be expected to be subjected to an overstressed condition. Ultimately, the members will crack or spall.

(j) Shrinkage :

- Shrinkage occurs due to moisture loss from concrete. It can be divided into two general categories: one that occurs before setting (plastic shrinkage) and one that occurs after setting (drying shrinkage).

- During the period between placing and setting, most concrete will exhibit bleeding of some degree. Bleeding is the presence of moisture on the surface of concrete; it is caused by the settling components of the mixture. Usually, bleed water evaporates slowly from the concrete surface. If the environmental conditions are such that evaporation is occurring rapidly than the water is being supplied from the surface by blooding, high tensile stress may develop. These stresses can cause cracks on the concrete surface.

(k) Temperature changes :

- The change in temperature causes a corresponding change in the volume of the concrete. As it was true for moisture-induced volume changes (drying shrinkage), temperature-induced volume changes must be combined with restraint before damage occurs. There are three temperature change events that can damage concrete. First, internal changes arise from the heat of the hydration of cement in a large placements. Second, there are temperature changes arising due to variation in climatic conditions. Finally, there is a special case of externally generated temperature changes - fire damage.

(l) Weathering :

- Weathering is often referred to as the cause of concrete deterioration. Weathering is defined as "changes in color, texture, strength, chemical composition, or other properties of a natural or artificial material due to weathering action". However, these effects may be more correctly attributed to other causes of concrete deterioration. Weathering is not considered as a specific cause of deterioration.

(m) Earthquake :

- An earthquake is a serious structural hazard that causes vibrations in structures due to ground shaking. These can collapse if the structures are weak, or extremely rigid to withstand severe vibrations. Tall buildings may experience extreme vibration due to their height, and may fall down into each other. Other devastating effects on structures due to earthquakes are sliding away from their foundations, and their horizontal or vertical movements can make structures unsafe.

(n) Termite :

- Common signs of termite infestation include sagging floors and ceilings, traces of dust similar to dust, piles of wings that resemble scales and areas that seem to be slightly water damaged. Termites are not only affect wooden structures but they have been known to inhabit new buildings within a short time after construction.

(o) Damage :

- Structural failure due to termite damage is very common, but it can be easily prevented through the use of regular inspections and treatments. Trusted termite inspectors will provide two-part reports, outlining damage already present and potential causes and locations of future damage. Areas with a high likelihood of infestation include damp areas, woodpiles and loose wooden paneling. Addressing these threats may prevent termite infestation and can save homeowners considerable money on structural repairs.

(p) Dilapidation :

- A building is dilapidated when it has undergone severe damage due to decay of different parts or members, mainly the structural members, causing serious imbalance in load transfer system. Due to decay the external appearance of the structure also becomes severely affected giving an unpleasant ugly look.

(q) Dampness :

- Dampness tends to cause secondary damage to a building. The unwanted moisture enables the growth of various fungi in wood or corrosion in steel causing rot or mould health issues and may eventually lead to sick building syndrome.

(r) Wind :

- Extreme wind events such as typhoons and tornadoes can cause devastating damage to structures and huge losses to human societies. Damage caused by wind can be minimized if the principles of fluid mechanics are applied at the design stage.

(s) Flood :

- The main damage from flooding comes from the impact of rapid moving waters. Flooding pushes the sides of structure, weakening it and at times separating it from foundation. Additionally, flooding penetrates deep into the earth causing severe damage

2.1.2 Classification of Failure of Structure

The failure of structure may be categorized as Structural and Non-structural failure.

Structural Failure :

Design failure	Constructional failure
<ul><li>Failure due to improper reinforcement.</li><li>Improper mix proportion.</li><li>Design mix.</li><li>Misconception of structural action.</li><li>Reinforcement detail.</li><li>Diffusion of concentrated loading.</li><li>Restrained shrinkage and temperature failure.</li><li>Standard for live load and wind loading.</li><li>Frequency of live loading. (Floor load, impact and inertia load, crane and lift load, hydrostatic or earth pressure).</li><li>Other loadings/design allowance for other loading.</li><li>Design provisions for wind loading.</li><li>Reinforcement grade and type.</li><li>Cover specified.</li><li>Design of life structure.</li><li>Shrinkage of concrete and relative movement between concrete and other materials.</li></ul>	<ul><li>Concrete- changes of concrete strength.</li><li>Accidental changes of concrete strength.</li><li>Use of admixtures not initially specified.</li><li>Reinforcement-substation.</li><li>Inaccurate spacing/placing.</li><li>Displacement.</li><li>Splices not as designed/joints.</li><li>Wrong reinforcement.</li><li>Omission.</li><li>Constructional joints - Expansion and Contraction joint.</li><li>Changes from specified procedure</li><li>Transportation, placing, compaction, finishing.</li><li>Stripping time (removal of covering)</li><li>Formwork removal</li><li>Re-shoring</li><li>Pre-stressing sequences.</li><li>Supervision - Amount of supervision.</li><li>Status of supervisor.</li><li>Person reasonable for supervisions.</li><li>Availability of construction records.</li></ul>

Non-Structural Failure :

Flood, fire, chemical attack.

Defective workmanship.

Temperature variation (hot and cold)

Design allowance for other loading :

- Construction loading.
- Differential shrinkage.
- Differential temperature.
- Foundation settlement.
- Differential vertical movement for creep.
- Blast, flood, earthquake, sever wind.
- Swelling soils.

2.2 SYSTEMATIC APPROACH OF DAMAGES DETECTION

2.2.1 Scope

- Prior to the commencement of visual inspection, the structural engineer is to obtain a set of the building's structural layout plans from the building owner.
- The availability of the structural layout plan will help the structural engineer to:

  (a) understand the structural system and layout of the building;

  (b) identify critical areas for inspection;

  (c) identify the allowable imposed loads, in order to assess the usage and possibility of overloading

  (d) verify if unauthorised addition or alteration works that affect the structure of the building have been carried out.

  The structural engineer is expected to carry out, with reasonable diligence, a visual inspection of :

  o   the condition of the structure of the building to identify,

    - the types of structural defects.
    - any signs of structural distress and deformation.
    - any signs of material deterioration.

  o   the loading on the structure of the building,

    - to identify any deviation from intended use, misuse and abuse which can result in overloading.

  o   any addition or alteration works affecting the structure of the building,

    - to identify any addition or alteration works which can result in overloading or adverse effects on the structure.

- If there are no signs of any structural deterioration or defects, the visual inspection should suffice and unless the structural engineer otherwise advises, no further action needs to be taken. If, on the other hand, signs of significant structural deterioration or defects are present, the structural engineer should make a professional assessment of the deterioration or defect and recommend appropriate actions to be taken. Such actions may involve full structural investigation to Parts or Whole repair work of the building.

2.2.2 Objectives of Investigation

- "Find the cause, the remedy will suggest itself". Sometimes, the source of the cause of distress is different than what is apparently seen. It is, therefore, essential that, the engineers conducting condition survey, determine the source(s) of cause so as to effectively deal with it and minimize their effects by proper treatment.

The Objectives of Condition Survey of a Building Structure :

- To identify causes of distress and their sources.
- To assess the extent of distress occurred due to corrosion, fire, earthquake or any other reason.
- To assess the residual strength of the structure.
- To assess its potential rehabilitation.
- To prioritize the distressed elements according to seriousness for repairs.
- To evaluate the influence of deterioration on serviceability if repair or replacement is deferred, and identify whether proposed changes to the structure would affect safety and service-life of the structure or not.
- To estimate the most cost-effective means to prevent further deterioration to the structure.
- To specify type; cause; and extent of deterioration, the rate of deterioration, and whether deterioration is active or not.
- To estimate the remaining time before repair or replacement.
- To estimate the most cost-effective means to prevent further deterioration to the structure.
- To study the performance of a material under a specific exposure condition.

2.3 INSPECTION TYPE

Condition based :	**Load based inspection :**
<ul><li>Location of structure.</li><li>Type of construction.</li><li>Design rick level.</li><li>Exposure condition.</li><li>Foundation condition.</li></ul>	<ul><li>Load depends upon the structure.</li><li>Durability of structure.</li><li>Over loading (ex: stairs, landing, passage, exit)</li><li>Appropriate loads on structure.</li></ul>
Periodic based inspection :	**Failure based inspection :**
<ul><li>It is the time based inspection.</li><li>Lift and escalators are periodically inspected or services.</li><li>Sanitary and plumbing works.</li><li>Record works are maintained periodically.</li><li>Removal of vegetation's.</li><li>Drainage and sewer lines are checked.</li></ul>	<ul><li>Construction error and non-clear supervision.</li><li>Improperly inspection construction joints.</li><li>Poor constructional material used and materials are not in standards.</li><li>Monitoring the important component of structure such as beam, column, roofs, stairs, etc.</li></ul>
Use based inspection :	
<ul><li>Type of structure.</li><li>Purpose of structure.</li><li>Appropriate use.</li><li>Periodic maintenance for structure will be used.</li><li>Serviceability and function ability based inspection.</li></ul>	

2.3.1 Planned Inspection

Preliminary Inspection

- The preliminary inspection is the first and most important single effort in the evaluation of a structure before rehabilitation. The preliminary inspection provides the initial analytical data that are used to assess the structural adequacy of an existing structure.

- Typical preliminary inspection involves the following phases:

1. **Gather Relevant Information :**

- Before conducting a preliminary inspection, the inspector normally reviews the available plans, specifications and construction records. Not only do these documents expedite the site inspection but also facilitate an evaluation that is more accurate and easier to perform.

- Some of the data items that are useful in most investigations are - name, location, type and size of the structure, date constructed, project engineer, contractor who built it, concrete supplier, materials used and sources; and test reports. In addition to code requirements that were applicable at the time of design, previous inspection reports, sketch map, photographs, exposure conditions, and changes to the structure since it was constructed are assessed.

2. **Identify Problems :**

- Specify whether the structure is safe to use or not. If the structure appears to pose a safety hazard, defer the inspection. Continue the inspection only after correction of the unsafe or hazardous conditions. Record physical damage or deterioration, displacement or misalignment of elements, foundation settlement, joint movement, staining, and dampness. Additionally, the effects of humidity: condensation, steel corrosion, *sulfate attack*, and pop outs should also be recorded.

- **The inspection should be made on the following points :**

  - Condition of Structural members
  - Structural defects like cracks, settlement, and deflection (sagging).
  - Condition of wall and wall paint.
  - Condition of flooring.
  - Roof leakage, leakage if any.
  - Dampness in wall or floor.
  - Condition of services fittings.
  - Drainage from terrace or pitched roof.
  - Growth of vegetation if any.

3. **Determine Possible Causes**

- It is extremely important that, the causes of a problem are properly identified in order to achieve a successful repair. The major causes of concrete problems are the environment, construction practices, materials, structural movement and previous repairs.

4. **Classification of Identified Problems :**

- Problems found in the structure are classified on the basis of the area affected, the depth affected, the volume of concrete affected, the width of cracks, the total length of cracks in a given area, or other relevant factors. Identified problems are classified into : light, medium, severe, very severe.

5. **Record Problems and Related Data :**

- During the preliminary inspection, the inspector records the nature and extent of the observed problems and identifies the affected members. Record the frequency and severity of the problems throughout the structure. If the location is accessible, inspect both the external and internal surfaces of a structure. If possible, photograph the areas of significant distress for later reference. Place a familiar object or scale in the photograph to show the relative size of the included area. Describe each observed problem in clear, concise detail.

- Use three-dimensional isometric drawings to show offsets or distortions of structural features, depth of delamination, location of reinforcement steel, and extent of steel corrosion.

6. Preliminary Inspection Report :

- Preliminary inspection report shall contain the following objectives :
    - Record the nature and extent of the observed problems.
    - Identify the affected members or areas.
    - Estimate or define the causes of the problems.
    - State the requirement for a detailed investigation.

7. Recommended Rehabilitation Method :

- If the preliminary inspection has identified the concrete problems, and a detailed investigation is not required, the report may also state the conclusions and make recommendations for the rehabilitation of the structure. Rehabilitation options can be divided into the following four categories: No repair, Cosmetic Repair, Structural Repair, Replacement.

2.3.2 Detailed Inspection

- On the recommendation of a structural engineer who carried out visual inspection, the BCA may approve for a complete structural investigation. If the structural deficiencies are of a localised nature, the structural engineer may recommend a complete structural investigation for that area in the first instance. The scope and extent of the investigation should be clearly defined and subject to approval. The result of this investigation may lead to a complete structural investigation for the entire building. The owner may engage a different structural engineer to complete this phase of the inspection and should notify government officials of the appointment.

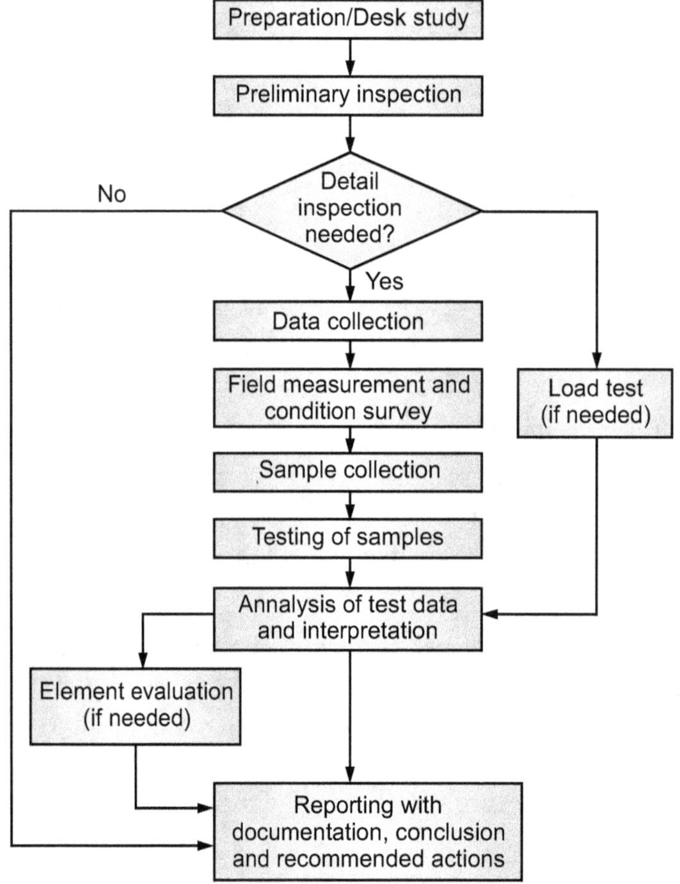

Fig. 2.1 : Methodology of inspection

Scope of Full Structural Investigation :

- Scope of a complete structural investigation includes the following:

  (a) Obtaining information related to the design, construction, maintenance and history of the building.

  (b) Assessing the structural adequacy of the building by examining structural plans and calculations and, reconstructing the structural plans if they are not available.

  (c) Carrying out tests on the materials used and structural elements used in the building.

  (d) Carrying out load tests on parts of the building, if necessary.

  (e) To recommend appropriate safety precautionary and remedial measures to restore the structural stability and integrity of the building structure.

- Field measurements and tests include delamination; voids; and crack mapping, corrosion activity/potential mapping, carbonation, concrete cover, spacing of rebars, concrete strength and variability, concrete quality (other than strength), and various non-destructive test methods.

- Samples are collected from the site to determine various properties of *in-situ* concrete. Samples are generally in the form of cores, broken concrete pieces, powdered samples extracted by drilling and sawed beams.

- Samples collected from the fields are tested in the laboratory to obtain necessary information which may include the compressive strength of concrete, cement content, chloride and sulfate content, permeability, aggregate type and gradation, alkali-aggregate reactivity, density and air voids.

- Inspectors should exercise caution in the interpretation of test results. The test results are easily misrepresented.

- **Specific Types of Detailed Inspections :**
  - Delamination Survey
  - Crack Survey
  - Half-Cell Potentials
  - Pachometer (Cover) Survey
  - Chloride Content
  - Moisture Content
  - Ultrasonic Testing
  - Rebound Hammer Test
  - Impact-Echo Test
  - Concrete Core Extraction and Test

2.3.3 Physical Inspection of Damaged Structure

The Steps in the Assessment Procedure to Evaluate Damages in a Structure :

- Physical inspection of damaged structure.
- Preparation and documenting the damages.
- Collection of samples and carrying out tests both *in situ* and in laboratory.
- Studying the documents including structural aspects.
- Estimation of loads acting on the structure.
- Estimation of environmental effects including soil structure interaction.
- Diagnosis.
- Taking preventive steps not to cause further damage.
- Retrospective analysis to get the diagnosis confirmed.
- Assessment of structural adequacy.
- Estimation on future use.

- Remedial measures necessary to strength and repairing the structure.
- Post repair evaluation through tests.
- Load test to study the behavior.
- Choice of course of action for the restoration of structure.

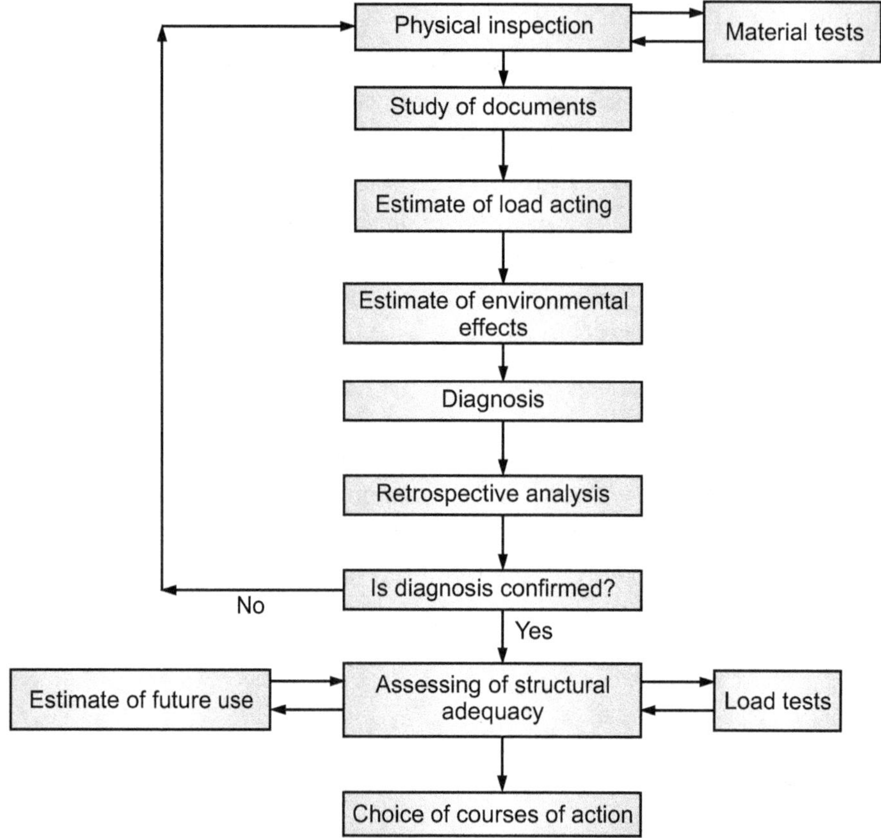

Fig. 2.2 : Determination of cause of damage

2.4 VISUAL OBSERVATIONS FOR DETECTION OF DAMAGES

- A key part of a structural evaluation is the visual inspection. It is also used to establish reference points and select sites for other tests. The comparison of test results from apparently sound areas with those from the damaged areas will generally provide much useful information in later analysis. Exposure conditions, the prevailing weather patterns and construction details such as overhangs, ledges and joints are all considered together when analysing test results.

- In addition to noting surface anomalies such as spalling, stains, cracks, delamination, blisters, popouts, pitting, efflorescence (fury or scaly deposits), the cracks size also should be measured, using a calibrated pocket microscope. Any deposits or gel in the cracks will also be seen with this device.

- One of the difficulties facing the engineer in carrying out the preliminary inspection on a large structure is that of an access. Where parts of the structure are beyond the reach of normal mobile platforms and scaffolds, such surveys are carried out solely by observation through binoculars and skilled technical photographic records. In spite of limitations, these methods are delightfully and deligently useful and reasonably effective. Large number of agencies and universities all over the world are currently working on the development of robotic devices that would carry out number of other sampling procedures and non-destructive testing along with video-camera records of surface details.

- What is often perplexing is to see severe damage only in certain parts of the building while other areas appear to be unaffected and in reasonably sound condition. This variable and localized damage is a phenomenon of interest and therefore attending this localized damage is relatively easy and permits to carry out localized repairs effectively. This controls restoration costs and duration of the activity.

2.4.1 Cracking of Concrete

- Cracking affects the appearance of concrete. In some cases it affects its structural adequacy and durability. In reinforced concrete, cracking allows easier access to air and moisture which can cause steel to rust and eventually weaken the concrete. The cracks can occur in two stages.

(a) Before concrete hardening :

- Movement of concrete causes these cracks to occur before the concrete has set. If they are detected before the concrete sets they can be retrowelled or refloated. If they are detected only when the concrete hardens, they should be filled with epoxy resin.

Plastic shrinkage cracks

- These type of cracks are encountered in hot and windy conditions. They appear as straight lines in parallel or as patterns. They are similar to crazing but on a larger scale. To avoid this type of cracking, the concrete should be compacted and finished as quickly as possible on hot days and the forms should be dampened to protect the concrete from heat and air.

Plastic settlement cracks

- This type appears while the concrete is still plastic and follows the lines of reinforcement.

Formwork movement cracks

- These cracks can occur during placement and compaction due to the movement of weak formwork. It is necessary to ensure that the formwork is strong and well propped and braced.

(b) After concrete hardening :

- Cracks, which occur in concrete when it hardens, may be due to drying shrinkage, settlement, structural cracks etc. They may require structural repairs such as high pressure epoxy injection or other means.

2.4.2 Spalling

- Spalling occurs when the edge or other surface of the concrete breaks. Spalling can be repaired by breaking out to sound and dense concrete then wetting and refilling the area with a cement material, which is then compacted, finished and cured.

2.4.3 Honeycombing

- When too much coarse aggregate appears on the surface with some cavities underneath is called honeycombing of concrete. This results in poor compaction or the mix is used with not enough sand. If it only occurs on surface it can be rearranged with a thin layer of mortar or a proprietary cement product. If cavities are present below the surface, it is more appropriate to break up the sound and dense concrete and repair it according to spalling above.

2.4.4 Dusting

- It is a surface defect that appears as a fine powder on the concrete surface and comes off when brushed. It is caused by finishing the concrete before bleed water has dried out, as well as by inadequate curing. It is repaired by applying a chemical floor hardener or a bonded topping.

2.4.5 Crazing

- This type of cracking resembles a map pattern. The cracks extend only through the surface layer. This is due to minor surface shrinkage as a result of drying conditions. It is avoided by finishing and curing as soon as possible. These cracks do not cause any subsequent deterioration of the concrete. If appearance is a problem, surface coating of paint can be applied to cover the cracks.

2.4.6 Rain damage

- Heavy rains can result in concrete piling or erosion of the surface. It is avoided by covering newly placed concrete with plastic sheeting when it rains. If there is rain damage and the concrete is not hardened, it can be reworked and refinished.

2.4.7 Efflorescence

- It is a white crystalline deposit, which appears immediately after completion. It is removed by dry brushing and flushing with clean water. Efflorescence has no effect on the structural performance of concrete.

2.4.8 Blistering

- Blisters occur when the fresh concrete surface is sealed by trowelling trapping air or bleed water under the surface. It is avoided to delay trowelling as long as possible and covering it to prevent evaporation.

2.4.9 Corrosion of Reinforcing Bars

- Corrosion occurs when the concrete surface cracks allows water penetration, or if water enters by diffusion during carbonation. The increase in the diameter of reinforcing bars due to formation of iron oxide (rust) can cause concrete above the affected bars to spall off.

2.4.10 Settlement

- If settlement has caused cracking and it is accordingly established, more complex devices i.e. extensometers or inclinometers may be used to monitor the movement over time period to determine if it is on-going, before an effective repair strategy is proposed.

2.4.11 Age of the Structure and Air-entrained Concrete

- Age of the structure is a common factor related to corrosion damage. Air-entrained concrete is recent provision for durability. Concrete exposed to freezing temperatures while wet, shall contain appropriately provided entrained air, Freeze-thaw damage in non-air entrained concrete gradually opens passage-ways by which moisture can reach the steel and corrode it.

2.4.12 Orientation of the Component of Structure

- Severity of exposure to prevailing wind and rain may make difference in damage levels between other exposed surfaces. South-West surfaces invariably manifest defects due to various reasons, into the cracking, discoloration, spalling, corrosion exposed reinforcement etc. These area surfaces prone to severe weather attacks need required additional restoration measures including protective coatings etc.

2.4.13 Insufficient Concrete Cover for Steel

- The most important factor for corrosion is insufficient cover of concrete causing direct damage to the RC element. Adequate cover and quality of cover concrete are of principal importance to resist the access of water and air, causing corrosion of steel in the concrete. Reinforcement placement errors or design deficiencies may be responsible for lack of adequate cover. Architectural provisions, cut-outs, inserts occasionally reduce the proper concrete cover to bars and hence due allowance must be made in the design. Variation in cover may cause galvanic cells to develop due to variations in available moistures, chloride ion or oxygen. Detailed survey of several structures clearly indicates that inadequate cover is one of the most important cause affecting concrete durability. The restoration measures must ensure adequate and good quality cover of concrete for long term effect of repairs.

- Low permeability is the most important property of the concrete in corrosion resistance. Thorough curing and low water-cement ratio are important in achieving low permeability. Although concrete dries slowly after moist curing is finished, the surface region dries most rapidly. Continuous curing is a must to avoid drying of cover concrete or also the major restoration problems hinge around achieving good quality cover concrete along with corrosion of reinforcement.

2.4.14 Significant Amounts of Soluble Chloride Ion in the Concrete

- Soluble chlorides contribute to the corrosion process wherever sufficient moisture and oxygen are present and this depends on the service conditions of the concrete. Extent of presence of chlorides in concrete would decide severity of the problems. Excess chlorides are difficult to remove during repairs, unless elaborated surface preparations are considered. These include major replacement of cover concrete, even substantially behind the reinforcement bars, which may attract structural instability. Also other proposals of cathodic protection may have to be conceived, if the structure is subjected to on-going severe aggressive environment. Extent of chloride penetration into the body of concrete also has to be considered. Effective and reasonably lasting repairs are not possible unless the chloride ingress is effectively dealt with.

2.4.15 Failure of Previous Repairs

- Inadequate or improperly applied patching materials may fail, or new deterioration may start just at the edge of the patch. All the modern bonding techniques and chemicals need strict compliance to the guidelines by the manufacturers. Proper surface preparation and timely of repairs, when fail, point out inadequate and sequential execution without mix-up. Most surface preparation, left over corrosion damage and suspect attended with the original surface is a weak spot and quality cover concrete. The contact line attracts weather effects and helps to activate damaging process further. The question therefore remains to be answered, how effective and lasting the repairs would be. Therefore one should be careful in projecting the commitments of age of repairs. Preventive measures and protection, if included, extends performance by few more years. For the structures which have received good repairs with technical soundness and quality awareness, 10 to 15 years of reasonable service life with regular routine protective/ preventive maintenance can be assumed reasonable.

2.5 TEST ON DAMAGED STRUCTURE

Strain Gauges :

- These may be electrical or mechanical gadgets and are used to monitor relatively finer and slow movement across or around cracks as load or thermal stresses change. They can therefore provide 'spot' measurements, or record a history of movement over a time period, when coupled to a suitable data-logging system.

Acoustic Emissions :

- Change in load or temperature induces stresses in a structure that can result in minute movements of the structural material around areas such as cracks, voids or intrusions. These movements generate acoustic noise that can be monitored and recorded. Analysis of the data can characterize each noise source signature by amplitude for a given stress and frequency content.

- Once the baseline signature for a point on the structure is established, changes in the signature either in frequency content or in amplitude, will indicate a corresponding change in structural condition, such as an increase in crack length, width etc.

Tell-tale Plates :

- The simplest crack monitors are thin glass plates glued across the crack. Any crack movement will break the glass. A more sophisticated unit consists of two acrylic plates. One is etched with a fine grid pattern, the other with a simple cross hair marking. The plates are glued to the either side of the crack, the cross hairs overlapping the grid and centered on it. Any subsequent movement of the sides of the crack can be measured from the position of the cross-hairs on the grid.

2.5.1 Importance and Need of Non-Destructive Testing

- It is often necessary to test concrete structures after the concrete has hardened to determine whether the structure is suitable for its designed use. Ideally such testing should be performed without damaging the concrete. The tests available for testing concrete range from completely non-destructive, where there is no damage to the concrete, through those where there is slightly damage to the concrete surface, to partially destructive tests, such as core tests, and pull-out and pull off tests, where the surface has to be repaired after testing. The range of properties to be evaluated using non-destructive tests and partially destructive tests is quite large and includes basic properties such as density, elastic modulus and strength as well as surface hardness and surface absorption, and reinforcement location, size and distance from the surface. In some cases, it is also possible to check the quality of workmanship and structural integrity by the ability to detect voids, cracking and roughness.

- Non-destructive testing can be applied to both old and new structures. For new structures, major applications are likely to be for quality control or to address doubts about the quality of materials or construction. Testing of existing structures is usually concerned with the evaluation of structural integrity or adequacy. In either case, if destructive testing alone is used, for example, by removing the core for compression testing, the cost of coring and testing may only allow a relatively small number of tests to be carried out on a large structure which may be misleading. Non-destructive testing can be used in those situations as a preliminary to subsequent coring.

2.5.2 Typical Situations where Non-destructive Testing May be Useful

- Quality control of pre-cast units or *in situ* construction.

- Removing uncertainties about the acceptability of the supplied material owing to apparent non-compliance with specification.

- Confirming or ignoring doubts about the workmanship involved in batching, mixing, placing, compacting or curing of concrete.

- Monitoring of strength development in relation to removal of formwork, completion of curing, prestressing, load application or similar purpose.

- Location and determination of the extent of cracks, voids, honeycombing and similar defects within a concrete structure.

- Determination of concrete uniformity, possibly preliminary to core cutting, load testing or other more expensive or disruptive tests.

- Determination of position, quantity or condition of reinforcement.

- Increasing the confidence level of smaller number of destructive tests.

- Determining the extent of concrete variability to help in the selection of sample locations representative of quality to evaluate.

- Confirmation or locating detection of suspected deterioration of concrete resulting from factors such as overloading, fatigue, external or internal chemical attack or alteration, fire, explosion, environmental effects.

- Assessing the potential durability of concrete.

- Monitoring long-term changes in concrete properties.

- To provide information for any proposed change of use of a structure for insurance or for change of ownership.

2.5.3 Situations where NDT is an Option to Consider for Investigation of *in situ* Concrete

- To investigate the homogeneity of concrete mixing.

- Lack of grout in post tensioning ducts.

- To determine the density and strength of concrete in a structure.

- To determine the location of reinforcing bars and the cover over the bars.

- To determine the number and size/diameter of reinforcing bars.

- To determine the extent of defects such as corrosion.

- To determine the location of in-built wiring, piping, ducting etc.

- To determine whether internal defects such as voids, cracks, delaminations, honeycombing, lack of bonding with reinforcing bars etc. exist in concrete.

- To determine if there is a bond between epoxy bonded steel plates and concrete members.

2.5.4 Basic Methods for NDT of Concrete Structures

The following methods, with some typical applications, have been used for the NDT of concrete :

- Visual inspection, which is an essential precursor to any intended non-destructive test. An experienced civil or structural engineer may be able to establish the possible cause(s) of damage to a concrete structure and hence identify which of the various NDT methods available could be most useful for any further investigation of the problem.

- A number of non-destructive, partially destructive and destructive techniques for assessment of concrete structure and to predict the cause of deterioration of the concrete in the existing structures are available. Interest in the field of Non-Destructive Testing (NDT) of structure is increasing worldwide. These NDT techniques can be broadly classified into following four groups:

1. **Strength Tests**	2. **Durability Tests**
• Schmidt Hammer Test • Ultrasonic Pulse Velocity • Pull out and Pull off Tests • Break off • Core Test • Windsor Probe • Pulse Eco Technique	• Corrosion Tests • Absorption and Permeability • Test for Alkali Aggregate Reaction • Abrasion Resistance Tests • Rebar Locator Test
3. **Chemical Tests**	4. **Performance and Integrity Tests**
• Carbonation Test • Suphate Determination Test • Chloride Determination Test • Thermoluminescence Test • Thermo gravimetric Analysis Test • Differential Thermal Analysis • Dilatometric Test	• Infrared Thermography Test • Radar Test • Radiography and Radiometry Tests • Acoustic Emission • Optical Fibre Test • Impact Echo Tests • Load Testing Test • Dynamic Response • X-Ray Diffraction

2.6 PHYSICAL TEST

2.6.1 Rebound Hammer Test

- The rebound hammer test is performed to assess the relative strength and elasticity of concrete on site based on its hardness at or near its exposed surface. Depending on the age of the concrete structure and carbonation effect, some specialized investigation is suggested before conducting the test. It essentially consists of a metal plunger, one end of which is held against the surface of the concrete while the free end is struck by a spring-loaded mass, which rebound to a point on a graduated scale. The point is indicated by an index rider. The amount of rebound increases with an increase in the strength of concrete for a particular concrete mix. It measures the surface hardness of the concrete and provides an estimation of surface compressive strength, uniformity, and quality of the concrete. User expertise is low and can be easily operated by field personnel.

- This gives an accurate assessment of the strength of the surface layer of the material. The entire structure can be tested in its 'as-built' condition. This can be very expensive and time consuming as instrumentation is required to measure the response. This requires careful planning and can damage the structure. The member must be separated from the rest of the structure before testing.

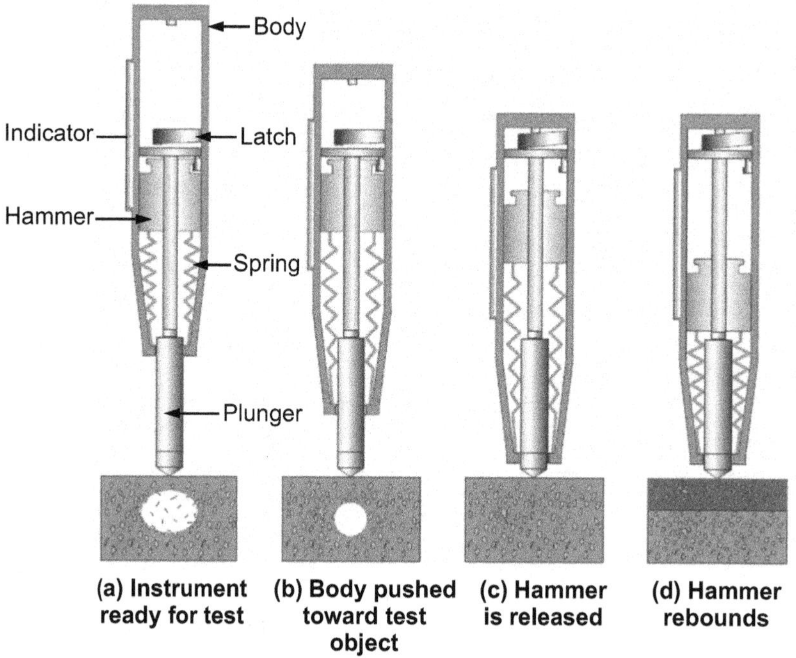

Fig. 2.3 : Rebound hammer test

- Although the rebound hammer provides a quick, inexpensive method of checking the uniformity of concrete, it has some serious limitations. Results are affected by:

1. Smoothness of test surface

- The hammer should be used against a smooth surface, preferably a formed one. Therefore, the open textured concrete cannot be tested. If the surface is rough, e.g. a trowelled surface, it should be rubbed smooth with a carborundum stone.

2. Size, shape and rigidity of specimen

- If the concrete does not form part of a large mass, any movement due to the impact of the hammer will result in a reduction in rebound number. In such cases the member should be rigidly held or backed up by a heavy mass.

3. Age of specimen

- For the same strength, higher rebound numbers are obtained with 7 days old concrete than with 28 days old. Therefore, when older concrete is to be tested in the structure, a direct correlation is necessary between the rebound numbers and compressive strength of cores taken from the structure. Rebound testing should not be carried out on low strength concrete at early ages or when the concrete strength is less than 7 MPa since the concrete surface could be damaged by the hammer.

4. Surface and internal moisture condition of concrete

- The rebound number is less for well-cured air dried specimens than for same specimen tested after being soaked in water and tested in saturated surface dry conditions. Therefore, whenever the actual moisture conditions of the field concrete or specimen is unknown, the surface must be pre-saturated for several hours before testing. A correlation curve for tests performed on saturated surface dried specimens should then be used to estimate the compressive strength.

5. Type of coarse aggregate

- Even though the same aggregate type is used in concrete mixing, the correlation curves can vary if the source of aggregate is different. There can be considerable differences between correlation curves developed for different aggregate types.

6. Type of cement

- High alumina cement may have a compressive strength 100% higher than the predicted strength using a correlation curve based on ordinary Portland cement. In addition, super sulfated cement concrete can have 50% less strength than ordinary Portland cement.

7. Carbonation of solid surface

- In older concrete the carbonation depth can be several millimeters thick and, in extreme cases, up to 20 mm thick. In such cases the rebound numbers can be 50% higher than those obtained on an uncarbonated concrete surface.

- The rebound reading on the indicator scale has been calibrated by the manufacture of the rebound hammer for horizontal impact.

Average Rebound Number	Quality of Concrete
> 40	Very good hard layer
30 to 40	Good layer
20 to 30	Fair
< 20	Poor concrete
0	Delaminated

2.6.2 Ultrasonic Pulse Velocity (UPV) Test

- An ultrasonic pulse velocity test is conducted as per IS 13111 - 1992 to assess the quality of concrete, in which the concrete under test is suspected to have low compaction, voids (porosity), delamination or damaged material.

- The ultrasonic pulse velocity test can also be used for the following applications:
  - Estimation of strength of concrete.
  - Establishing homogeneity of concrete.
  - Studies on durability of concrete.
  - Analysis of surface crack depth.
  - Measurement of changes occurring with time in the properties of concrete.
  - Correlation of pulse velocity and strength as a measure of concrete quality.
  - Determination of the modulus of elasticity and dynamic Poisson's ratio of the concrete.

- Voltage pulses are generated and transformed into wave bursts of mechanical energy by the transmitting transducer (which must be coupled to the specimen surface through a suitable medium). A receiving transducer is coupled to the specimen at a known distance to measure the interval between transmission and reception of a pulse. There are three practical arrangements for measuring pulse velocity, namely direct, diagonal and surface techniques. The direct approach provides the greatest sensitivity and is therefore superior to other arrangements. Determination of variability and quality of concrete by measuring pulse velocity. By using the transmission method, the extent of such defects such as voids, honeycombing, cracks and segregation can be determined. This technique is also useful when examining fire damaged concrete. Low levels are required to make measurements. However, expertise is needed to interpret the results. Excellent for determining the quality and uniformity of concrete. It can rapidly survey large areas and thick members. Path lengths of 10 m to 15 m can be inspected with appropriate equipment.

- Proper surface preparation is required. The work takes a lot of time because it only takes point measurements. Skill is required in the analysis of results as moisture variations and presence of metal reinforcement can affect result. Interpretation of ultrasonic test results based on published graphs and tables can be misleading. It is therefore necessary that correlation is observed with concrete. It works on single homogenous material.

- The receiving transducer detects the arrival of that component of the pulse, which arrives earliest. This is usually the leading edge of longitudinal vibration. Although the direction in which the maximum energy propagates is at right angles to the face of the transmitting transducer, it is possible to detect pulses, which have traveled through concrete in another direction. Therefore, it is possible to measure pulse velocity by placing two transducers on either:

  - Opposite faces (direct transmission)

  - Adjacent faces (semi-direct transmission): or

  - The same face (indirect or surface transmission).

UPV value in km/sec (V)	Concrete quality
V greater than 4.0	Very good
V between 3.5 and 4.0	Good, but may be porous
V between 3.0 and 3.5	Poor
V between 2.5 and 3.0	Very poor
V between 2.0 and 2.5	Very poor and low integrity
V Less than 2.0 and reading fluctuating	No integrity, large voids suspected

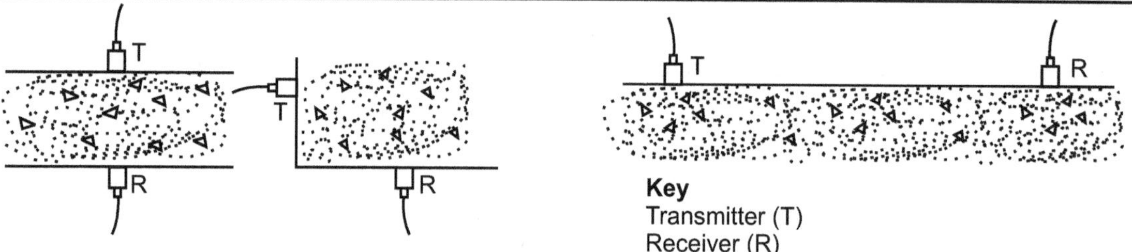

Fig. 2.4 : Various positions of key in ultrasonic pulse velocity test

Sr. No.	Test Results	Interpretations
1.	High UPV values, high rebound number	Not corrosion prone.
2.	Medium range UPV values, low rebound numbers	Surface delamination, low quality of surface concrete, corrosion prone.
3.	Low UPV, high rebound numbers	Not corrosion prone, however, to be confirmed by chemical tests, carbonation, pH.
4.	Low UPV values, low rebound numbers	Corrosion prone, requires chemical and electrochemical tests.

2.6.3 Rebar Locator/Cover Gauge

- By this test, bar diameter, cover to reinforcement, spacing of reinforcement, number of reinforcement bars and any discontinuity in the reinforcing bars can be detected. This test is performed using a cover meter which is based on the electro-magnetic principle. A rebar locator is used to determine the presence and orientation of steel reinforcement rebars under the concrete surface. A contractor engaged in maintenance work will be familiar with the problem of accurately locating the exact position of the rebar, wall ties, studs and other metal fasteners. These low cost, simple to use gauges can meet their everyday requirements.

- Test hammers are used to determine the surface hardness of concrete and are one of the most widely used instruments for assessing the compressive strength of concrete. It is the quickest, simplest and least expensive method for obtaining estimates of the quality and strength of concrete. Test hammers with both analog and digital displays are available. Many concrete structures have a protective or cosmetic coating. Premature failure of this coating can, at the very least, result in additional cost of rework.

- Adhesion tests verify that both the surface preparation and coating application are within specification. Concrete structures are porous and will absorb moisture, our range of moisture meters and climate monitoring gauges allow moisture content to be measured. The more extensive range includes gauges used for measurement of crack width in concrete and other structures.

Principles and Procedures:

- The reinforcement bar is detected by magnetizing it and inducing a circulating "eddy current" in it. After the end of the pulse, the eddy current dies away, creating a weak magnetic field as the echo of the initial pulse. The strength of the induced field is measured by a search head as it dies away and this signal is processed to give the depth measurement. The eddy currents echo is determined by the depth of the bar, the size of bar and the orientation of the bar. This detection of location of reinforcement is required as a pre-process for core cutting.

Reliability and limitations:

- With this instrument, the reinforcement cover can be measured with instruments up to 100 mm with an accuracy of ± 15% and manufacturers claim the accuracy of bar diameter measurements with less than 2 to 3 mm. But it has been observed that most of the available detectors do not measure the diameter correctly. Proper calibration of these instruments is very essential.

Factors affecting accuracy are :

- Very closely spaced bars or bundled bars or bars in layers, binding wire, aggregate containing iron or magnetic properties.

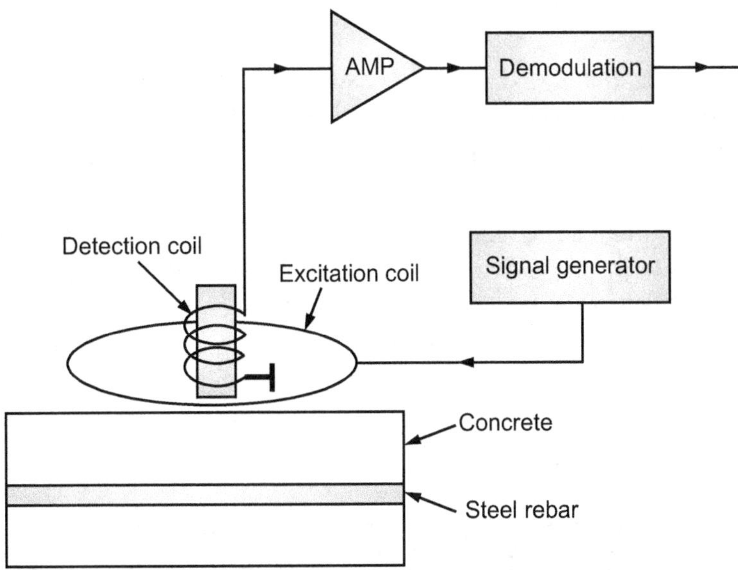

Fig. 2.5 : Rebar locator

2.6.4 Crack Detection Microscope/Digital Crack Measuring Gauge

- The digital position strain gauge deformation meter consists of a digital dial gauge attached to a bar. A fixed conical point is mounted at one end of the bar, and a moving conical point is mounted on a knife edge pivot at the opposite end. A setting out bar is used to position the pre-drilled discs which attached to the structure using a suitable adhesive. Whenever readings are to be taken, the conical points of the gauge are inserted into the hole in the disc and the reading on the dial gauge noted. In this way, the strain changes in the structure are converted into changes in the readings on the dial gauge.

- The gauge has been designed so that only a minor temperature correction is necessary for changes in ambient temperature, and a reference bar is provided for this purpose.

  Measuring range : Compression - 1.6 mm, Tension - 3.4 mm.

  Minimum measuring unit : 0.001 mm.

  Standard accessories: Digital crack gauge, Calibration bar, Installation bar, Locating discs (box of 100).

Crack Magnifier

- Crack widths are normally limited to 0.2 mm or 0.3 mm in concrete structures. This crack measuring device enables accurate determination of, whether cracks exceed this limit.

  Magnification : 10x.

  Measuring range : 20 mm × 0.1 mm.

  Field of view : 32 mm.

  Special design permitting to read on light and dark objects.

Crack Width Meter

- The crack width meter is used as a comparator to give an approximate crack size during visual surveys.

  Graduations from 0.1 mm to 2.5 mm.

Crack Monitor

- On some structures the rotation at cracks is also significant. This gauge is specifically designed to measure rotation, transverse and longitudinal movement. The crack monitor is used for :

  Measuring movements: Horizontal ± 20 mm.

  Reading accuracy of: ± 0.5 mm on grid.

Crack Meter

- The Vibrating Wire Crack Meter are used to measure movement across surface cracks and joints in concrete, rock, soil and structures. They consist of a sensor outer body tube and an inner free sliding rod which is connected at the internal end to a vibrating wire sensor by a spring.

Field Microscope

- The field microscope is a small sized lightweight and conveniently portable microscope. Designed to cover the range between high grade heavily equipped microscopes and measuring magnifiers. With a magnification of 50 times, this microscope can be used to accurately measure the width of cracks and also combines a calibrated focusing ring allowing the depth of cracks to also be accurately measured.

  Magnification : 50x.

  Measuring range : 1.6 mm × 0.02 mm.

  Field of view : 1.7 mm.

  Comes with light.

2.7 PARTIAL DESTRUCTIVE TESTS

2.7.1 Pullout Test

- This test measures the force required to pull out a specifically shaped steel rod whose enlarged end has been embedded in concrete. The force is measured with the help of a dynamometer. The pull out strength is calculated from the ratio of the pull out force to the idealized area of frustum of the cone. Concrete strength is determined with calibrated correlation between pullout strength and compressive strength.

There are two options for the pullout test :

- o   DANISH LOK TEST which requires that the head to be cast into concrete at the time of construction. This test gives a good indication of near surface compressive strength.

- o   Building Research Establishment, UK (BRE) PULLOUT involves drilling a hole and inserting a "fixing" which is pulled out. The advantage of this test is that it does not require a head to be cast into the concrete during construction. The disadvantage is that the test actually measures tensile strength and is then calibrated to compressive strength.

- A pullout test is a test that falls into the transition zone between a destructive test and a non-destructive test. It is destructive in the sense that a relatively large volume of concrete is damaged but non-destructive because the damaged can be repaired.

- The pullout test measures the force required to pull an embedded metal insert with an enlarged head from a structure. The insert is pulled by a loading ram seated on a bearing ring that is concentric with the insert shaft. The bearing ring transmits the reaction force to the concrete. Frustum geometry is controlled by the inner diameter of the bearing ring (D), the diameter of the insert head (D), and the embedment depth (H).

- The pullout tests are widely used during construction to estimate the in-place strength of concrete to determine whether critical activities such as removal, application of post tensioning, or termination of cold weather protection can proceed. Since compressive strength is usually required to evaluate structural safety, the ultimate pullout load measured during in-place test is converted to an equivalent compressive strength by means of a previously established correlation relationship.

- As the insert is pulled out, a conical shaped fragment of concrete is extracted from the concrete mass.

- Unlike some other tests that used to estimate the in-place strength of concrete, the pullout test subjects the concrete to a slowly applied load and measures an actual strength property of the concrete. However, the concrete is subjected to a complex three dimensional state of stress, and the pullout strength is not likely to be related simply to uniaxial strength properties. Nevertheless, by use of correlation curves the pullout test can be used to make reliable estimates of in-place strength. An important step in implementing the method is choosing the locations and number of pullout tests in a given placement of concrete. The inserts should be located in the most critical portions of the structure and there should be a sufficient number of tests to provide statistically significant results. Additional inserts are recommended in the event that testing begins too soon, and the concrete has not attained the required strength.

- The use of maturity meters along with the pullout tests is encouraged to assist in selecting the correct testing times and in interpreting possible low strength results. The BRE pullout test was developed to permit testing in an existing construction by drilling a hole and inserting some type of expansion anchor. The results of these tests are difficult to interpret if a correlation curve does not exist for the concrete used in the construction.

2.7.2 Pull Off Test

- This test involves attaching a plate to the concrete using epoxy resin and, after curing has taken place, measures the force required to pull the plate off. This test scares the concrete, but gives a measurement of the near-surface tensile strength that can be converted to compressive strength provided a correlation exists between compressive strength and tensile strength for the concrete mix being investigated.

2.7.3 Core Test

- This is a direct way to assess the strength of concrete. In this method, cylindrical core samples are taken from existing structures. The core is inspected and tested in the laboratory to check its compressive strength.

- This test is used to determine the compressive strength of the concrete core, which is usually extracted from the existing structure. The value of compressive strength can then be used in conjunction with other measured properties to assess the condition of the concrete.

- Using a masonry saw, the core is first trimmed to the correct test length, which varies if the standard is adopted. After trimming, the core will either have a completely flat ground, or be capped into the material to produce a smooth bearing surface.

- After the prescribed curing has taken place, the specimen is then crushed to failure noting the maximum load achieved. From the values of load and dimensions, the compressive strength of the core can be calculated.

- In most structural investigations or diagnoses, extraction of core samples is unavoidable and often essential. Cores are usually extracted by drilling using a diamond tipped core cutter cooled with water. Broken samples, for example, due to popping, spalling and delimitation, are also usually retrieved for further analysis because these samples may provide additional evidence as to the cause of the distress. The selection of locations for the extraction of core samples is made after non-destructive testing which can give guidance on the most suitable sampling areas.

- For instance, a cover meter may be used to ensure that there are no reinforcement bars where the core is to be taken; Or ultrasonic pulse velocity tests can be used to establish areas of maximum and minimum pulse velocity that can indicate the highest and lowest compressive strength areas in the structure.

- Moreover, by using non-destructive tests, the number of cores that are required to be taken can be reduced or minimized. This is often an advantage since coring is frequently viewed as being destructive. Also the cost of extracting cores is quite high and the damage to the concrete is severe.

- The extracted cores can be subjected to a series of tests and serve multiple functions such as:
  - confirming the findings of the non-destructive test.
  - identifying the presence of deleterious matter in the concrete.
  - ascertaining the strength of the concrete for design purposes.
  - predicting the potential durability of the concrete.
  - confirming the mix composition of the concrete for dispute resolution.
  - Determining specific properties of the concrete not attainable by non-destructive methods such as intrinsic permeability.

2.8 CHEMICAL TEST

2.8.1 Chemical Analysis

- The chemical composition of concrete as determined from extracted core samples is usually essential in making an informed assessment of the concrete condition, its potential durability and suitability for continued use. These include, for instance, chloride content, sulphate content and pH value. The analyses are often requested for evaluation for compliance against the specification, in particular, the mix composition as compared to the design. Findings obtained from the analyses can then be used, for instance, to formulate a repair programme and resolve disputes.

- Typical **information obtainable and usually required include the following:**
  - cement content
  - original water content and water cement ratio
  - aggregate grading
  - chloride content
  - sulphate content.

- Sometimes an analysis of the type and amount of admixtures is requested. Such an analysis is much more complicated and difficult, as it requires special techniques and instruments. Examples include gas liquid chromatography, high pressure liquid chromatography, Fourier transformed infrared spectroscopy and thin layer chromatography. Other allied techniques to identify various characteristics of the hardened concrete include X-ray fluorescence spectroscopy, differential thermal methods and X-ray diffraction. As in all attempts to break down the composition of a complex material, there are bound to be uncertainties.

- Ideally, the individual constituents used in concrete should be made available for analysis. However, practically, this is rarely possible usually because these ingredients are no longer available at the time of testing. This is especially the case for old structures. In such situations, assumptions have to be made with regard, for example, to the types of cement and aggregates used and typical composition of the cement mixed.

- Adequate sampling is probably one of the primary factors in the uncertainties of the analysis particularly when considering the heterogeneity of concrete in large structures. The assumptions made in the analysis in the absence of available raw materials would also contribute to errors. Subsequent secondary reaction or degradation of the concrete in service such as carbonation, leaching, chemical attack and alkali silica reaction can also induce inaccuracies in the analysis.

2.8.2 Chloride Test

- Small amount of chlorides will normally be present in the concrete. Higher amount of chlorides may give rise to potential of corrosion risk. Quantity of chlorides in the concrete is generally determined chemically and is expressed in terms of percentage of chlorides by weight of concretes.

- Thermo gravimetric and Dilatometric test, differential thermal analysis tests, Thermolumine scene test etc. are some of the sophisticated tests for assessment of the residual concrete strength.

2.8.3 Sulphate Attack

- National codes of different countries give different limits for the aggressive sulfate content for the safe service of concrete structures. The international organization CEMBUREAU set 600 mg 50427L (600 ppm) in water and 6000 mg S042-/kg (6000 ppm) in soil as the boundary between moderate and severely aggressive sulfate media where sulfate resisting cement should be used. Based on these limits of sulfate content, the European Prestandard ENV 206 states 500 ppm in water and 3000 ppm in soil as its criteria (Lawrence 1990). However, ACI recommended 1500 ppm S042- in water and 2000 ppm 5042- in soil respectively. Usually, more detailed recommendations are made, e.g. such as that summarized by BRE for concrete in service in sulfate-containing soil and ground water.

2.8.4 pH Measurement

- Although concrete typically begins its life at a high basic pH of about 13, the pH value on exposed surfaces soon drops as reactions occur between carbon dioxide from the atmosphere and alkali in the concrete - a process is called carbonation. Over time, fronts of carbonated concrete, with a pH value of about 8.5, advance below exposed surfaces. It may be necessary to determine the depth of this front, as carbonated concrete may allow corrosion of the reinforcing steel. This can be done by spraying phenolphthalein indicator solution on a fractured or cut surface of concrete and focusing on the color change location.

- In this case, a color change occurs when the solution contacts a material with a pH of 9 to 9.5. The depth of the carbonation front can also be evaluated using a drill to collect samples at the selected depth. In this case, the pH of the powdered samples can be evaluated using standard test methods.

Field Test for pH Measurement of a Concrete Surface :

- Clean the surface using a wire brush. Be sure to remove dirt, concrete sealer, and old adhesive residue;
- Gather 0.5 g (0.018 oz) of concrete powder by hand sanding an approximately 50 × 50 mm (2 × 2 inch) area with 50-grit general-purpose sandpaper;
- Thoroughly mix the concrete powder with 10 to 12 drops of fresh distilled water with a small, flat plastic stirrer. Let stand for 60 seconds;
- Insert a pH strip into the mixture. Compare the strip to the color chart to determine the pH; and note the temperature of the concrete surface.

Laboratory Test for pH Measurement of Concrete :

- Calibrate the pH probe and meter using the manufacturer's directions and appropriate buffer solutions;

- Collect 5 g (0.18 oz) of concrete powder at the required depth using a drill;

- Use a plastic stirrer to mix the concrete powder with 10 mL (0.34 fl oz) of fresh distilled water at a temperature of 22 ± 1 °C (72 ± 2 °F);

- Filter the mixture using No. 40 filter paper;

- Insert the pH probe into the mixture. Read the pH to one decimal place; and note the temperature of the mixture.

2.8.5 Resistivity Method

- The electrical resistance of concrete plays an important role in determining the quality of concrete from the point of view 'corrosion susceptibility potential' at any specific location. This parameter is expressed in terms of "Resistivity" in ohm-cm.

- For general monitoring, a resistivity check is important because long-term corrosion can be anticipated in concrete structures where accurately measured values are below 10,000 ohm-cm. Further, if resistivity values fall below 5,000 ohm-cm, corrosion must be anticipated at a much earlier period (possibly within 5 years) in the life of a structure. Table indicates the general guidelines of resistivity values based on which areas having probable corrosion risk

Sr. No.	Test Results	Interpretaion
1.	High resistivity greater than 10,000 ohm-cm and low potentials – more positive than –200 mV (CSE)	No active corrosion relatively cathodic.
2.	Low resistivity below 10,000 ohm-cm and potentials between – 200 mV to –350 mV (CSE)	Initiation of corrosion activity – relatively anodic.
3.	Low resistivity about 5,000 ohm-cm and potentials 200 mV to –350 mV (CSE)	Presence of corrosion potential – activity – anodic.
4.	Low resistivity below 5,000 ohm cm and negative than – 350 mV (CSE)	High intensity of potential more corrosion – fully anodic
5.	Higher potential gradient and high conductivity.	High rate of corrosion

Interpretation about Corrosion Based on Chemical Analysis of Concrete :

The following Table gives some tips for interpretation about corrosion :

Test Results from Chemical Analysis	Interpretation about Corrosion
High pH value (> 11.5) and very low chloride content	No corrosion
High pH value and high chloride content greater than 0.4% - 0.6% by weight of cement	Corrosion prone
Low pH value and high chloride content greater than 0.4% - 0.6% by weight of cement	Highly corrosion prone

2.8.6 Half-Cell Potentiometer Test

- Electrochemical half-cell potentiometer testing provides a relatively quick way to assess reinforcement corrosion over a wide area without the need for bulk removal of the concrete cover. The method of half-cell potential measurements normally involves measuring the capacity of an embedded reinforcing bar relative to a reference half-cell placed on a concrete surface. A half-cell is usually a copper/copper sulphate or silver/silver chloride cell but other combinations are used. The concrete functions as an electrolyte and the risk of corrosion of the reinforcement in the immediate region of the test location can be empirically related to the measured potential difference. In some circumstances, useful measurements can be obtained between two half-cells on the concrete surface. ASTM C876 - 91 gives a standard test method for half-cell potentials of Uncoated Reinforcing Steel in concrete. Quantitative measurements are made so that structure can be monitored over a period of time and deterioration can be noted. Areas of use include marine structures, bridge decks and abutments and so on. Used in combination with other tests, it has been found to be helpful when investigating concrete contaminated by salts.

Principle and Procedure :

- The instrument measures the potential and the electrical resistance between the reinforcement and the surface to evaluate the corrosion activity as well as the actual condition of the cover layer during testing. The electrical activity of the steel reinforcement and the concrete leads them to be considered as one half of weak battery cell with the steel acting as one electrode and the concrete as the electrolyte. The name Half Cell Survey derives from the fact that one half of the battery cell is considered to be a steel reinforcing bar and surrounding concrete.

- The electrical potential of a point on the surface of steel reinforcing bar can be measured comparing its potential with that of copper – copper sulphate reference electrode on the surface. Practically this is achieved by connecting a wire from one terminal of a voltmeter to the reinforcement and another wire to the copper sulphate reference electrode. Then generally, readings taken are at grid of 1×1 m for slabs, walls and at 0.5 m c/c for column, beams.

- The risk of corrosion is evaluated by means of the potential gradient obtained; higher the gradient higher is the risk of corrosion.

- The test results can be interpreted based on following table :

Half-cell potential (mv) relative to Cu-Cu sulphate (Ref. electrode)	% chance of corrosion activity
Less than – 200	10%
Between – 200 to – 350	50% (uncertain)
Above – 350	90%

Significance and Use:

- This method may be used to indicate the corrosion activity associated with steel embedded in concrete.

- This method can be applied to members regardless of their size or the depth of concrete cover. This method can be used at any time during the life of concrete member.

Reliability and Limitations:

- The test does not indicates actual corrosion rate or whether corrosion activity has already started, but it indicates the probability of the corrosion activity depending upon the actual surrounding conditions. If this method is used in combination with resistivity measurement, the accuracy is higher.

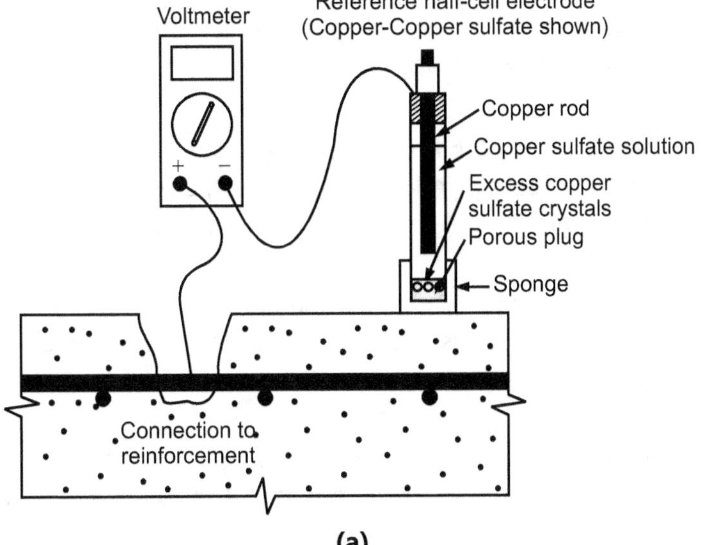

(a)

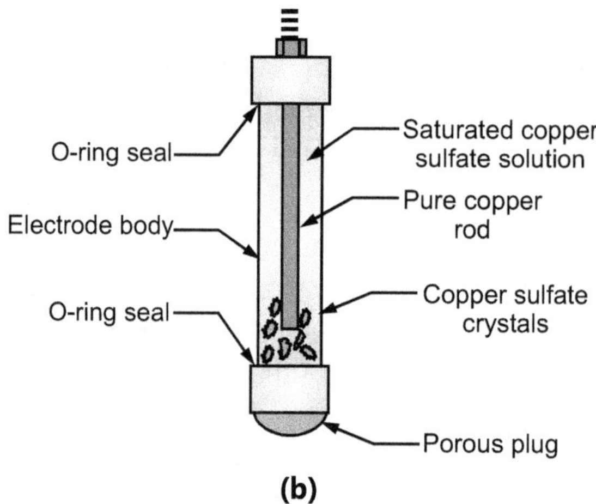

Fig. 2.6 : Half-cell potentiometer test

- If the concrete surface has dried to the extent that it is dielectric, then pre wetting of concrete is essential especially for Cement Silos, Exposed roof slab. The quality of the cover concrete, particularly its moisture condition and contamination by carbonation and/or chlorides may affect the results.

2.8.5 Carbonation Test

- The method of testing involves the determination of the depth of carbonated layer on the surface of hardened concrete by means of an indicator. Carbonation of concrete occurs when carbon dioxide, in the atmosphere in the presence of moisture, reacts with hydrated cement minerals to produce carbonate, e.g. calcium carbonate. The carbonation process is also called depassivation.

- Carbonation penetrates very slowly below the exposed surface of the concrete. The importance of carbonation is that the usual protection of reinforcing steel usually present in concrete is due to the alkaline conditions caused by the hydrated cement paste, which is neutralized by carbonation. Thus, if the entire concrete covering the reinforcing steel is carbonated, corrosion of the steel will occur if moisture and oxygen can reach the steel. The time required for carbonation can be estimated knowing the concrete grade and using the equation.

- Permeability of concrete is important when dealing with the durability of concrete. Concrete durability largely depends on the ease or difficulty with which fluids (water, carbon dioxide, oxygen in the form of liquid or gas) can drain through hardened concrete mass. This particularly is applicable in water retaining structures or watertight substructures which are exposed to harsh environmental conditions and require low porosity as well as permeability. Such adverse elements can result in degradation of reinforced concrete. The permeability test measures the ease with which liquid, ions, and gases movement can occur by flow, diffusion, and absorption. Generally, the overall potential for moisture and ion ingress in concrete by these three modes referred to as its permeability.

Important Points

- **The main causes of deterioration of concrete structures** are design and construction flaws, Poor quality material used, Accidental loading, Environmental effect, Corrosion of embedded metals. Abrasion, Cavitation, Freezing and thawing, Settlement and movement, Shrinkage, Temperature changes. Weathering, Earthquake, Termite.

- "Find the cause, the remedy will suggest itself". Sometimes, the source of the cause of distress is different than what is apparently seen. It is, therefore essential that, the engineers conducting condition survey, determine the source(s) of cause so as to effectively deal with it and minimize their effects by proper treatment.

- The preliminary inspection is the first and most important single effort in the evaluation of a structure before rehabilitation. The preliminary inspection provides the initial analytical data that are used to assess the structural adequacy of an existing structure.

- On the recommendation of a structural engineer who carried out visual inspection, the authorities may approve for a complete structural investigation. If the structural deficiencies are of a localised nature, the structural engineer may recommend a complete structural investigation for that area in the first instance.

- Visual inspection is used to establish reference points and select sites for other tests. The comparison of test results from apparently sound areas with those from the damaged areas will generally provide much useful information in later analysis.

- Cracking affects the appearance of concrete. In some cases, it affects its structural adequacy and durability. In reinforced concrete cracking allows easier access to air and moisture which can cause steel to rust and eventually weaken the concrete. The cracks can occur in two stages.

- Spalling occurs when the edge or other surface of the concrete breaks. Spalling can be repaired by breaking out to sound and dense concrete then wetting and refilling the area with a cement material, which is then compacted, finished and cured.

- The rebound hammer test is performed to assess the relative strength and elasticity of concrete on site based on its hardness at or near its exposed surface. Depending on the age of the concrete structure and carbonation effect some specialized investigation is suggested before conducting the test.

- Ultrasonic Pulse Velocity (UPV) Test is conducted to assess the quality of concrete, in which the concrete under test is suspected to have low compaction, voids (porosity), delamination or damaged material.

- By Rebar locator test, bar diameter, cover to reinforcement, spacing of reinforcement, number of reinforcement bars and any discontinuity in the reinforcing bars can be detected. This test is performed using a cover meter which is based on the electro-magnetic principle. A rebar locator is used to determine the presence and orientation of steel reinforcement rebars under the concrete surface.

- Pullout test measures the force required to pull out a specifically shaped steel rod whose enlarged end has been embedded in concrete. The force is measured with the help of a dynamometer. The pull out strength is calculated from the ratio of the pull out force to the idealized area of frustum of the cone. Concrete strength is determined with calibrated correlation between pullout strength and compressive strength.

- **Pull off test** involves attaching a plate to the concrete using epoxy resin and, after curing has taken place, measures the force required to pull the plate off.

- **Core test :** This is a direct way to assess the strength of concrete. In this method, cylindrical core samples are taken from existing structures. The core is inspected and tested in the laboratory to check its compressive strength.

- The chemical composition of concrete as determined from extracted core samples is usually essential in making an informed assessment of the concrete condition.

- Higher amount of chlorides may give rise to potential of corrosion risk. Quantity of chlorides in the concrete is generally determined chemically and is expressed in terms of percentage of chlorides by weight of concretes.

- Although concrete typically begins its life at a high basic pH of about 13, the pH value on exposed surfaces soon drops as reactions occur between carbon dioxide from the atmosphere and alkali in the concrete - a process is called carbonation.

- The electrical resistance of concrete plays an important role in determining the quality of concrete from the point of view 'corrosion susceptibility potential' at any specific location. This parameter is expressed in terms of "Resistivity" in ohm-cm.

- **Half-Cell Potentiometer Test** provides a relatively quick way to assess reinforcement corrosion over a wide area without the need for bulk removal of the concrete cover. This method may be used to indicate the corrosion activity associated with steel embedded in concrete.

- **Carbonation Test** involves the determination of the depth of carbonated layer on the surface of hardened concrete by means of an indicator.

Questions for Practice

1. Explain the causes of deterioration of concrete structures ?
2. Give classification of failure of structure.
3. State the scope for damage detection?
4. Write the objectives of Condition Survey of a building structure?
5. Classify and explain types of Inspection.
6. Explain the relevant procedure of repairing the damage occurred in the building.
7. Enlist the parameters to decide damages in the building.
8. Give the relevance of non-destructive technique to decide the severity of damage of the given structural element?
9. State various ND tests performed on concrete.
10. Explain the chemical analysis of concrete.
11. State typical situations where non-destructive testing may be useful.

❑❑❑

$\mathscr{Chapter}$ **3**

MATERIALS FOR MAINTENANCE AND REPAIRS

Weightage of Marks = 14, Teaching Hours = 10

Syllabus

3.1 Factors influencing the material selection for maintenance and repairs.

3.2 Anti-corrosion coating materials : Cement slurry mortar, polymer modified cement slurry and epoxy zinc.

3.3 Adhesive materials : Solvent free adhesives : epoxy adhesive, polyester adhesive, acrylic adhesive and water borne adhesives : polyvinyl acetate and vinyl acetate co-polymer.

3.4 Mortar repair materials : Cementitous mortar, polymer modified cementitious mortar and resin mortar.

3.5 Joint sealants materials : Oleo resinous mastics, bitumen/rubber based sealants and acrylic resin sealant.

3.6 Grout materials : Cement grout, cement sand grout, cement sand grout with additives, polymer modified cement grout and normal expoxies.

3.7 Waterproofing roof materials : Polyisobutylene (PIP) sheet, glass fiber reinforced plastics, bitumen and bituminous emulsion and latex cement coating.

3.8 Surface coating materials for concrete protection : Bituminous cutbacks, chlorinated rubber coating, vinyl coatings, epoxy coating and coal tar epoxy.

3.9 Additional repairing materials : plastic or aluminium nipples, polyester putty or 1 : 3 cement sand mortar and galvanized steel wire fabrics and clamping rods.

Objectives

After learning this chapter, student will be able to -

- Select the relevant materials to repair the given type of damages with justification.

- Explain the suitable materials for the anti-corrosion, adhesives and mortar repairs.

- Explain the relevant materials for waterproofing and joint sealants for the repairing of given structure.

- Explain the relevant materials for surfacing coating and grout for the repairing of given structure.

- Choose the relevant materials for the repair of damaged structures with justification.

INTRODUCTION

- The selection of repair materials is one of the most important tasks to ensure durable and reliable repairs. Pre-requisite for a sound repair system is detailed investigation and determining the exact cause of the distress. But understanding of the deterioration process of repair materials under service conditions is important. The relevance, availability of equipment and materials of skilled labor must be ascertained before deciding on a repair.

3.1 FACTORS AFFECTING THE SELECTION OF MATERIAL FOR REPAIR

- The analogy for the selection of repair materials is similar to repairing a torn garment with sound fiber/fabric, but with the same performance characteristics as the original garment. If the patch repair was done un-shrunk or with dissimilar fiber/fabric, the damaged garment would have suffered more damage due to its shrinking after washing. In fact, it applies to the selection of materials for repair of concrete/plaster. Also the selection of repair materials has a chemical angle and the manufacturer's literature generally sheds light on the structure of the material rather than on the performance characteristics. Since cement products have a tendency to shrink and harden with age, it is essential that the repair material for concrete or plaster repair must be of a non-shrinking type and be compatible with the original material.

- **Important factors which influence the selection of materials for repair** are as follows:

    1. Bonding of repair materials to existing substrates.
    2. Non-shrink properties of repair material (volumetric stability).
    3. Strength of repair materials as compared to old materials of construction.
    4. Resistance to shrinkage crack formations.
    5. Type and extent of damage.
    6. Type of structure.
    7. Environmental conditions.
    8. Serviceability conditions.
    9. Site conditions.
    10. Time factor.
    11. Temperature at application.
    12. Maintenance of proper cover.
    13. Durability of repair material.
    14. Curing of newly laid repair materials.
    15. Matching of repair materials with substrate.

3.1.1 Essential Parameters for Maintenance and Repair Materials

- In addition to the compatible properties, repair materials for cement concrete/mortar will also be easy to apply and will require no attention after repairs are applied. The necessary parameters for deciding on the repair material for concrete are :

(a) Low shrinkage :

- It is well known that cementitious repair materials shrink with the passage of time. Most of the shrinkage typically occurs in the initial period up to 21 days from the time of casting. Therefore, cementitious repair material in its original form, when used to repair concrete/mortar, is likely to disintegrate or develop cracks due to de-bonding on its surface due to shrinkage strains and stresses. Shrinkage cracks developed in repair patches will allow easy access to atmospheric air and water, which can be harmful to concrete and reinforcement.

- Therefore, it is necessary that the low shrinkage property of the repair material be looked for when selecting materials for repairing concrete. The cementious material needs additional non-shrinking compounds to be effective in achieving the desired property. Therefore, the manufacture of patch mortar incorporates several special chemicals into the cement matrix, reducing shrinkage. Using a lower cement content and a lower water-to-cement ratio will also reduce drying shrinkage.

(b) Requisite setting/hardening properties :

- It is desirable that the repaired structure be put to use as soon as possible to reduce the down time of plant, machinery, building or road. Therefore, it is necessary that, the repaired patches harden in the minimum possible time. However, in exceptional cases, it may also be necessary to have a slow setting property as a desirable property for repair materials. There may be a situation where more work time is required to work on the repair material or the repair process is complicated that more work time is required.

(c) Workability :

- The repair material has to be applied by the field workers and hence its acceptance by them is very important. Property desired by field workers is of good workability. Therefore optimum workability is to be achieved without sacrificing the other desirable properties by use of suitability additives/admixtures.

(d) Bond with the substrate :

- The bond strength of the repair patch with the substrate is essential for a successful repair system. If it is felt that the bond strength of the repair material with the base material is insufficient or less than the strength of the base material, then some other suitable means to improve the bond strength between the repair material and the substrate can be discovered is.

  They can be used :
  - Adhesive,
  - Surface Interlocking System, and/or
  - Mechanical Bonding

- Unmodified polymer applications are available in a range of adhesives, epoxies, including polymer modified cement slurries. The selection depends on the open time available for bonding etc., being specified for various applications. Surface interlocking systems and mechanical bonding methods are, however, detailed in a later chapter on repair methods.

(e) Compatible coefficient of thermal expansion :

- The difference in volume change due to temperature variation can cause failure either on the bonding line or in the section of low strength material. Therefore, in areas exposed to temperature variations, the repaired patch must have the same coefficient of thermal expansion to ensure that no undue stress is transferred to the bonding interface or substrate. Due to the same coefficient of thermal expansion, cemented materials are preferred over epoxy materials.

(f) Compatible mechanical properties and strength :

- The hardened materials will have compatible mechanical properties or slightly better strength than base materials. This property is desirable to ensure uniform flow of stresses and strains in loaded structures. It is well known that the elastic modulus of two concrete is different for different crushing strength, so if the repaired concrete has much higher strength than the base, it can lead to non-uniform flow of stress and this may result in an initial failure of repair patch. For example, if the M-20 grade of concrete has been used in the original construction, the grade of the repaired material shall not be less than M-20 or more than M-25.

(g) Relative movement, if required :

- Especially in the case of sealing cracks where movement is expected or at expansion joints, the selected repair material will be flexible and elastic, capable of absorbing the anticipated relative movements of the structure without any signs of distress or cracks.

(h) Minimal or no treatment required :

- It is desirable that, after the repair is applied, the repair material will not require any curing or, if it is necessary, it should be at a minimum to ensure that the repair patch hardens and achieves the desired strength without much post-repair care. Only epoxies do not require any curing. Other material applications require minor to moderate curing, which needs to be specified. There are such material have several other properties which must be taken into consideration when selecting required material. The curing compound can be applied to cement materials used for repair but after checking its compatibility.

(i) Low air and water permeability :

- The permeable material allows easy saturation of environmental chemicals including carbon dioxide, water, oxygen, industrial gases/vapors etc. It is essential that, the repair material must have very low air/water permeability to provide protection to reinforced concrete against the ingress of harmful environmental chemicals.

(j) Aesthetics :

- It is desirable that the color and texture of the repaired material should match with the structural and give an aesthetically pleasing appearance. If necessary, this can be achieved through suitable finishes.

(k) Cost :

- Economics is important when considering various options for repair materials but cheaper repair materials should not be selected at the cost of performance characteristics.

(l) Durability and Bio-non degradation :

- The selected repair material must be durable against its exposure condition during service life against chemical attack, resistant to any type of energy such as ultra violet rays, infra red rays, heat etc. and must be bio non-degradable.

(m) Non-hazardous/non-polluting :

- Repair material should not be dangerous for field workers. However, adequate safety measures need to be taken for repair materials, which are hazardous to the workers involved in their application etc. They should also be environmentally friendly.

3.1.2 Types Of Repair Materials

1. Anti-corrosion coatings.
2. Adhesives /Bonding aids.
3. Mortar repair materials.
4. Curing compounds.
5. Joints sealants.
6. Grout materials
7. Waterproofing roof materials.
8. Specials concretes.
9. Migratory corrosion inhabitor.
10. Protective aoatings.
11. Materials for special Applications.

3.2 ANTI-CORROSIVE COATING MATERIAL

- There are essentially two forms of inhibitors; those that can be admixed into fresh concrete and spread on the surface of the concrete to regenerate by an occasional application. An inhibitor can be defined as, "a substance that, when added in small amounts, decreases the metal corrosion rate in a potentially aggressive environment".

- Over the last half century. Many compounds have been tested as corrosion inhibitors for reinforced concrete, with varying degrees of success. Many of which have been employed successfully in other industries.

- Aqueous corrosion involves two synchronous reactions, the anodic reaction where the metal ions dissolve in the solution and the rest in the cathodic reaction where the hydroxyl ion is formed by mixing oxygen and water. Electrons released at the anode are consumed at the cathode and the resulting processes are described as electrochemical reactions. Only the anode suffers material loss, but both sites required for corrosion proceed and corrosion products form when reaction products accumulate at both the aligned and oxidized (rust) sites.

- There are three main types of inhibitors that work by suppressing one or both electrical reactions that cause erosion, namely anodic and cathodic reactions.

Anodic inhibitors : The anodic reaction is suppressed by passing the metal, forming a film on the surface, and preventing an anodic reaction from taking place. This category is further divided into oxidized and non-oxidized types. Oxidizing anodic inhibitors require the presence of oxygen to act and contain substances such as nitrite. Non-oxidizing anodic inhibitors do not require the presence of oxygen as substances such as phosphates, silicates, and polyphosphates.

Cathodic inhibitors : Typically, a film of salts is precipitated by reacting with anions in porous solutions such as hydroxyl at cathodic sites. Examples of cathodic inhibitors are divalent cations such as calcium, magnesium, zinc, strontium and chromate.

Mixed inhibitors : The inhibitor can react with both adsorbs on the entire metal surface (anodic and cathodic) and can react with corrosive product, for example, organic compounds such as amines. They are also called as ambiodic inhibitors.

- The inhibitors can be further distinguished by the mode of application as, Diffused-in or Admixture. Diffused-in inhibitors are applied to the surface of the concrete and drain through the pore structure of the concrete on the surface of the rebars. Admixed corrosion inhibitors are combined in fresh concrete at the time of mixing.

3.2.1 Characteristics of Ideal Anti-Corrosive Materials

- Resistance to entry of moisture by film formation.
- Resistance to under cutting (i.e. progressive rust creep under the primer).
- Good bond to steel and subsequent repair layers.
- Resist adverse effect on the adjacent steel.
- Easy to apply.

3.2.2 Types of Anti-corrosive Materials

1. Polymer modified cement slurry.
2. Cement mortar slurry.
3. Zinc rich epoxy.

1. Polymer Modified Cement Slurry :

- Polymer modified cementitious slurry coating is applied over the base concrete. This system can be used with new and old roofs. This technique of water proofing is the latest development in the field of water proofing. Polymer provides significant improvement in imperviousness. Polymer modified cemented slurry has the same expansion coefficient as concrete, which does not cause cracks due to the thermal variation of concrete.

- Polymer modified cemented slurry coatings consist of a liquid mixture and a dry mixture. Liquid mixtures contain polymers, liquid additives and clean water. Dry mixtures include locally available portland cement and aggregates. These mixtures are usually mixed in a ratio of 2 : 1 (cement : polymer) and applied by brush to the prepared surface on a weight basis. Generally, two coats are provided. However, the manufacturer's recommendations in this regard should be followed. To enforce greater flexibility, it is desirable to use some reinforcement such as a fiberglass cloth sandwich between two coats. This coating should be covered with screed concrete to avoid physical damage.

2   Cement Mortar Slurry (Corrosion Inhibitor-containing Cementitious Slurry Coating)

- Steel reinforcements which are embedded in concrete are surrounded by an alkaline medium. So, a coating based on cement is expected to be more suitable. A passivating cement coating may have higher tolerance towards the defects. Galvanic effect is likely to be less pronounced because the surrounding concrete is alkaline in nature.

- In view of economy and efficiency, Central Electrochemical Research Institute (CECRI), Karaikudi, India developed a coating based on Portland cement slurry admixed with special corrosion inhibitor. The coating is made impermeable to salts by sealing treatment. Inhibited cement slurry coating was also known as corrosion inhibitor-containing cementitious slurry coating.

3 Zinc Rich Epoxy :

- Zinc is an amphoteric metal which reacts with both strong acid and strong base solutions. The reaction will be more pronounced for pH value below 6 and pH value above 13. At intermediate pH ranges, the rate of attack on zinc is very slow due to the formation of protective surface layers. When embedded in concrete, zinc is passivated for pH values between 8 and 12.5, again due to the formation of a protective surface film of corrosion product which is insoluble below pH 12.5.
- Zinc reacts with wet cement, and the reaction ceases once the concrete has hardened and the barrier layer of calcium hydroxyzincate has formed. Hydrated concrete is strongly alkaline with a pH in excess of 12.2 due to the presence of a saturated solution of $Ca(OH)_2$ filling the pore space. Corrosion of steel in concrete is usually caused by either a natural reduction in the pH of concrete through reaction with acidic gases such as carbon dioxide. Corrosion may also occur due to carbonation effect, or due to the presence of chloride ions above certain threshold levels at the depth of the reinforcement.
- Steel in concrete typically depassivates below pH 11.5, or higher in the presence of chlorides, leading to the onset of corrosion. Zinc remains passivated at a pH of 9.5 thereby offering substantial protection against the effects of carbonation of the cover concrete. Zinc can also withstand exposure to chloride ion concentrations several times higher than that causes corrosion of black steel. The corrosion protection thus afforded by galvanizing is due to the combined effect of a substantially higher chloride threshold than black steel in concrete, and a complete resistance to the effect of carbonation of the concrete.

3.3 ADHESIVES

- Application to conventional concrete, sprayed concrete or sand-cement repair mortars, adhesion between materials is often a problem. When the repairs are to be carried out at a high temperature, water loss at the interface between the repair materials and the prepared concrete element surface may prevent proper hydration of hydraulic matrices along the interface resulting in poor bond. Hence under such conditions, it is necessary to provide a bonding coat prior to the application of repair material.
- The major advantage of adhesive bonding is that it allows distributions of an applied load over much larger area thus reducing the unit stress on the repaired elements. It allows adhesion without changing the shape of the element to be attached. The adhesive bond line also acts as moisture barrier. There are two main types of adhesives :
  (a) Solvent free adhesives and
  (b) Water borne adhesive (e.g. latex and powder)

(a) Solvent Free Adhesives (e.g. polymeric adhesives) :

(i) Epoxy adhesives :

- Epoxy adhesives are usually made of epoxy resin amine or polymer metal, which are reactive diluents and in some cases inorganic filler and thixotroping agents. They are the lowest polymer adhesives. Epoxy adhesives generally have excellent adhesion due to the relatively low curing shrinkage with low surface tension and molecular properties, which increases their attractiveness for a wide variety of substrates. They are very tolerant of the alkalinity of concrete.
- The ratio of resin to curing agent in most epoxy adhesives is very low, allowing for proper metering and mixing within the tolerances of available automated equipment. Epoxy adhesives that will cure and bond underwater are also available. Since resins systems (resin/curing agents) are available with viscosity less than 100 cps and in semi-solid form, they can be prepared in adhesive products that penetrate but in bond-lines or in a product contaminants are required which can fill gaps without being contained. Epoxy can be tailored to suit a variety of strength and module requirements. The water and chemical resistance of epoxy adhesives, after curing as a class, is excellent and can be compared to only polyester adhesives.

(ii) Polyester adhesives :

- Unsaturated polyester resins usually dissolve in the Styler mononer. They are cured with initiator, usually an organic peroxide. Promoters and accelerators are commonly used to activate the decomposition of the initiator at room temperature, thus enabling rapid low temperature curing. Due to their relatively high shrinkage when curing, polyesters have limited use as adhesives.

- Resistance to bond failure can be increased by increasing the flexibility of polyesters, thus relieving some local stresses during the application of external forces. Most polymers do not bond to moist or wet substrates and should not occur when these conditions are present.

- Recent research has noted well that some vinyl esters, a type of polyester, can bond under such conditions.

- Polyester curing can be accelerated by the addition of an accelerating component that can provide complete curing in about two minutes. Therefore mixing with only automatic equipment is desired. Polyesters generally have excellent resistance to acid environments. Some polyesters have relatively poor resistance to alkalis and solvents. Although the polymer's water resistance itself is good, most polyester adhesive bonds are continuously degraded in wet conditions.

- Polyesters are generally considered flammable with flash points below 38°C. A few additional explosions of inert filler to form a combination can reduce hazards. However, prolonged storage of the initiator is not appropriate as peroxide decomposition occurs.

(iii) Acrylic adhesive :

- Acrylic adhesive methyl methacrylate and high molecular weight methacrylate monomers of the acrylic family are used as solvent-free adhesives for concrete. These adhesives usually share the same characteristics as polyester adhesives. They are most commonly used by mixing with fine aggregates to create an easily flowing adhesive mortar.

- The flow capacity of the mortar can be controlled with the addition of aggregate amount. The cure time is between 30 minutes to 2 hours, which is considered sufficient in the field. In almost all cases, the primer is made of methacrylate monomer cured with organic peroxide and is used to improve the bonding of concrete.

(iv) Polysulfide adhesives :

- Polysulfide adhesives are most often used as flexibilisers in epoxy resin formulations. These aggregates are sometimes referred to as polysulfide adhesives and fit properly into the category of 'epoxy adhesives'. Polysulfide materials are mainly used as joint sealants.

(v) Polyurethane adhesives :

- These are used as both rigid and flexible materials. When combined with aromatic amines, the urethane forms a rigid polymer similar to epoxy adhesives. They have limited use with concrete due to their low bond strength. Flexible types are used as membrane systems and for ceramic tiles of concrete, where impact resistance is required.

(vi) Silicon adhesives :

- Silicones that have the ability to cure in a wide temperature range are almost exclusively used as sealants of flexible joints. However, they can be used to bond elements such as windows to concrete where highly flexible adhesives are required to reduce stress concentrations. Silicon should not be used in applications requiring resistance to constant load.

(b) Water Borne Adhesives :

- They are latex and latex powder adhesives. The only water-borne adhesives currently used for bond concrete are latex and latex powder adhesives. There are two types I and II.

- The type I is used without further formulation and the type II used in slurry form with hydraulic cement generally portland cement solution. For Type II adhesives, the ratio from latex to cement is about 1 : 4 by weight.

- Both types of adhesives are typically used to bond fresh, unhardened concrete to hardened concrete. However, Type II adhesives have also been used for some time to strengthen hardened concrete. Latex and latex powders are usually made by emulsion, polymerization techniques.

(1) In performance verification, when the surface is held, the reinforcing material must be bonded to the concrete by adhesive. The bond strength of concrete and reinforcing materials will vary depending on the type of adhesive and reinforcing material, the nature of the concrete surface and other factors, so suitable tests should be conducted to check the bond strength.

(2) The adhesive used for the overlap splices using steel splice plates and the lap splices for the continuous fiber sheets must have a bonding strength sufficient to enable the required overlap splice strength to be obtained.

(3) The fiber in a continuous fiber sheet must be well resinous between continuous fibers to interconnect and transmit stress, to enable the sheet to act as a fiber-reinforced plastic with Proper strength, Young's modulus and other quality requirements. Accordingly, the impregnation/adhesive agent acts as an adhesive that bonds the concrete to reinforcing material, as well as being able to ensure the strength, Young's modulus and other quality requirements for continuous fibre sheet as fibre bonding material for fibre reinforced Plastic.

(4) With the bonding method, the structure is retrofitted on site using adhesives to bond the concrete reinforcing material. Accordingly, the adhesive must have viscosity, fluidity, and other quality values appropriate for site construction.

- Type I and type II adhesives are generally used for bonding fresh unhardened concrete to hardened concrete. However, Type II adhesives have occasionally been used for bonding hardened concrete. Latex powder is usually made by emulsion polymerization techniques. These adhesives include the following :

(i) Polyvinyl acetate (PVA).

(ii) Vinyl acetate co-polymers (VAC).

(iii) Poly acrylic esters (PAE).

(iv) Styrene-butadiene Copolymers (SB).

(i) Polyvinyl Acetate :

- Polyvinyl acetate latexes are Type I adhesive and are usually formulated with a plasticizer such as dibutyl phthalate or di-propyl glycol di-benzoate. The plasticizers are added to decrease the minimum film forming temperature. This type of adhesive is usually made in polyvinyl alcohol surfactant system and is available in latex form and as a redispersible powder. Water resistance of such adhesives is suspect because of hydrolysis of the polyvinyl acetate. Films of the latex are redispersible.

(ii) Vinyl Acetate Copolymers :

- Copolymers of vinyl acetate with such materials as butyl acrylate, ethylene and the vinyl ester of versatic acid are Type I adhesives but can also be used as Type II adhesives. They are generally made in polyvinyl alcohol surfactant systems and are available in latex and redispersible powder forms. Their water resistance is much better than that of polyvinyl acetate, because the comononer reduces the hydrolysis of the vinyl acetate grouping and also the resultant product is not as water soluble as polyvinyl alcohol. The water resistance of such polymers will depend on the type and ratio of comonomer to vinyl acetate. The comonomer also causes a reduction in the minimum film forming temperature, which eliminates the need for addition of plasticizers. When used as type II adhesives, bond strengths usually exceed 6.9 MPa.

(iii) Polyacrylic Esters and Acrylic Co-Polymers :

- Polyacrylic ester latexes, such as polyethyl acrylate, and acrylic co-polymer latexes are Type II latex adhesives. They are generally made using primarily a non-ionic surfactant system. If the latex dries before placement of the fresh concrete, the dried film can act as a bond breaker rather than as an adhesive. These groups can improve adhesion by ionic reaction with metallic radicals in the surface of the fresh concrete. However it has been observed that such groups may retard the initial hydration of the hydraulic cement.

(iv) Styrene-butadiene Co-polymers :

- Styrene-butadiene copolymer latexes are Type II adhesives. They could be used as Type I adhesive but are not recommended for this category, because their films are not redispersible. In addition, their surfactant system is primarily of the non-ionic type. Such groups can improve adhesion and latex stability, but may also retard the initial hydration of the hydraulic cement.

3.3.1 Applications of Adhesives

- The adhesive used to bind the concrete and reinforcing material must be one that can ensure the required bonding strength.
- The adhesive used for overlap splices for reinforcing materials must be one that can ensure the strength of the overlap splice section.
- The impregnation/adhesive agent used for the continuous fiber sheet bonding method should be one that ensures the strength, Young's modulus, and other quality requirements of the continuous fiber sheet as the fiber bonding material.
- The adhesive must be of appropriate viscosity, shrinkage and other characteristics, including coating, fill or other fabrication method.

3.4 MORTAR REPAIR MATERIALS

- There are many reasons for the deterioration of the reinforced concrete elements and there various ways in which these can be repaired. Deterioration mechanisms caused by corrosion of reinforcement, sulphate attack or alkali-silica reaction, result in extensive cracking. In such situations, it is necessary to remove damaged concrete and refill it with repair mortar or concrete.

3.4.1 Applications of Mortar Repair Materials

- Material for surface preparation.
- Chemical corrosion removers for rusted reinforcement.
- Passivaters for reinforcement protection.
- Bonding agents.
- Structural repair materials.
- Non-structural repair materials.
- Injection grouts.
- Joint sealant.
- Surface coatings for RCC protection.

3.4.2 Types of Mortar Repair Materials

(a) Cementitious mortars/concrete.

(b) Polymer modified cementitious mortars.

(c) Resin mortars.

(a) Cementitious Mortars/Concretes :

(i) Conventional cement concrete/mortars.

(ii) Gunite/shotcrete.

(i) Conventional Cement Concretes/Mortars : Portland cement concrete or mortar offer a number of advantages as repair materials. Thermal properties are similar to existing concrete, similar in appearance and comparatively of low cost. It has ready availability and ease in applying. Concrete is often preferred for complete replacement of sections and deep cavities extending beyond reinforcing bars. Mortars can be used for small cavities/spalls (as small as 10-15 mm). Portland cement mortar with water content low enough to render it no-slump concrete may be used and referred to as 'dry pack'. It finds its application as a good repair material. Low water contents reduces shrinkage of the repaired portion while the effectiveness gets increased. The properties of cement mortar/concrete can be further modified with the addition of suitable admixtures like superplasticizers, hydrophobic agents, pozzolanas, fly ash, silica fume, volcanic ash, diatomaceous earth etc.

(ii) Gunite/Shotcrete : Gunite is a cementitious repair material that is pneumatically applied to concrete surfaces. Shotcrete or gunite is mortar or concrete conveyed through a pressure line and pneumatically at a high velocity on to a damaged worn out prepared surface. It has the advantages of requiring no formwork, self-compacting and monolithic joints. Using specially graded aggregates and equipment, 28-day equivalent cube strengths in excess of 50 N/ mm^2 can be obtained. Reinforced gunite can be obtained by pinning the reinforcement old surfaces and then guniting. Gunite linings are usually reinforced with steel bar fabric (welded mesh) and the distribution steel may be more than that provided in the original concrete. Normally, gunite would have a w/c ratio of about 0.5, through w/c ratio up to 0.33 can be used. Several types of fibres and additives are added to the gunites to improve its properties and performance.

(b) Polymer Modified Cementitious Mortars/Concretes :

- Since the early 1950s, it has been known that certain polymers can be added to cementitious mortars to help overcome many of the problems of using unmodified mortars or concrete as repair materials. The polymers used as admixtures for cementitious systems are normally supplied as milky-white dispersions (latex) in water and are used to gauge the cementitious mortar as a whole or, as partial replacement of mixing water. Such mortars offer the same alkaline passivation protection of steel as conventional cementitious materials. They can be placed in a single application of 12-15 mm thickness which gives adequate protective cover.

- **The advantages of using polymer latex are :**
  - It functions as a water-reducing plasticizer, producing a mortar with good.
  - Workability and lower shrinkage at lower water/cement ratios.
  - It improves the bond between the repair mortar and the concrete being repaired, provided it is applied and used properly.
  - It reduces the permeability of the repair mortar to water, carbon dioxide and oils and also increases its resistance to some chemicals.
  - It acts as an integral curing aid, but careful curing is generally essential.
  - It increases the tensile and flexural strength of the mortar.

- Styrene butadiene rubber (SBR), acrylic and modified acrylic latexes are most commonly used as modifiers in repair mortars and base concrete. When they are properly formulated for compatibility with cement, there are no significant differences in the long term performance of the repair mortars. The latex modified concrete has the advantage of being, able to be designed for adequate workability and placed in a manner similar to ordinary concrete by conventional methods. However, site batching and mixing often results in production of unsatisfactory mortars due to lack of adequate quality control. To overcome this problem, there are now complete prepacked sets of latex and preblended sand and cement available which simply require mixing at site.

- Polymer modified cementitious mortars are mainly used for the repair of reinforced concrete, where the cover to be replaced is more than 12 mm and less than 30 mm in thickness. Above 25-30 mm of cover, conventional cementitious repair materials are commonly used.

(c) Resin Mortar :

- Where the cover to be replaced is less than 12 mm and areas to be repaired are relatively small, resin mortars are being used. When resin mortars are used, the protection of steel reinforcement depends wholly upon the impermeability of the repair mortar. This requires very careful application including surface preparation of the steel reinforcement to a high standard. Resin mortars are based on reactive resins filled with carefully graded aggregates. Epoxy resins are most commonly used, but polyester and acrylic resins are also been used. Epoxies, polyester and acrylic resins are classified as thermosetting materials. When cured, the molecular chains are locked permanently together. They are generally supplied as 2-3 component system (resin, hardener and filler).

1. Epoxy Resins :

- Epoxy resins ranging from low molecular weight liquids to high molecular weight solids can be obtained. However, this type of basic resin is not suitable for many applications due to its high viscosity. Modification of basic resin is therefore necessary to achieve the required wetting capacity, curing rate and many other properties in addition to low viscosity. Epoxy resins are chemical mixtures with high tensile strength that serve as exceptional adhesive agents. Their composition can be changed as per requirement and the ingredients required to make the appropriate epoxy resin are mixed just before application. The viscosity of some epoxy resins is so low that they can also be injected into fine cracks. On the other hand, epoxy resins that have a high viscosity are typically used for surface coating purposes and to fill large cracks or holes. These binding agents can also be used effectively in applications that need to attach steel plates to damaged structures.

- Epoxy resin consists of a reactive resin which is cured with a curing agent called 'hardener'. Concrete proportioning and correct mixing are essential operations when using epoxy resin. The curing of epoxy resin system is an exothermic reaction and the rate of curing is temperature dependent. As a rule, the rate of curing doubles as the temperature increases by 10°C. Epoxy-based formulations possess high chemical resistance but low elasticity. These are also suitable for filling joints subjected to low movements. These are used to repair concrete or masonry damaged due to chemical attack area subjected to high impact or tensile loads, leakage of chemical tanks, cracks of dams and spillway, to fill cracks and honeycomb.

Characteristics of Epoxy Resin :

- **High strength :** Compressive strength of 60 to 80 N/mm^2 and very tensile and flexural strength (usually 5 to 10 times greater than concrete)
- **Thin sections :** Sections as thin as 5 mm can safely be applied.
- **Adhesion :** Exceptional adhesion to concrete; repaired area never fail along correctly prepared bond line.
- **Rapid cure :** Most of the strength gain occurs on first day in normal atmospheric condition.
- **Chemical resistance :** Excellent resistance to alkalis; very good resistance to most acids and solvents; excellent resistance to most of the minerals.
- **No maintenance :** Correctly applied repairs need no replacement.
- **Impermeable :** Properly compacted mortar is impermeable to water, water-borne contaminants and airborne gases.
- Most epoxy resin must be applied on dry surface. Ambient air temperature and relative humidity has to be kept within fairly narrow limits.

Limitations :

- These are not suitable above 100°C temperature.

   (i) Unsaturated reactive polymer resins : Polyester resin systems are chemically much more simple. Both the reactive components are present in the resin. The hardener is catalyst which is required only to initiate the reaction. Mixing and proportioning of hardener components is therefore, less critical than for epoxy resins. There is reduction in volume in the set polymer and so, polyester resin formulations must be limited to applications in relatively smaller areas at one time.

Advantages :

- **Very rapid cure :** Polyester develop strength extremely quickly; repaired areas can be reused within one to three hours.
- **Part mixing :** Parts packs may be used.
- **No priming :** Polyester tends to rely on mechanical rather than chemical bonding.
- Independent primer is not necessary. Though in specific condition priming may be helpful.

Limitations :

- Areas greater than approximately 30 sq. cm. will be prone to shrinkage.
- Repair sizes are limited.
- Deep repair need to be applied in layers in order to avoid heat built-up.

(ii) Unsaturated Acrylic Resins : Acrylic resin systems form high strength materials by chemical cure mechanism. In general, acrylic resins are based on monomers of very low viscosity or blends of monomers with methyl methacrylate monomer. Acrylic resins permit higher proportions of fillers because they have lower viscosity and so the mortars exhibit less shrinkage upon curing.

2. Epoxy mortar :

(i) Epoxy Mortar and Concrete :

- Epoxy resins are used for the manufacture of epoxy mortar or epoxy concrete with aggregates (silica sand) used for structural repair of concrete. RCC other than this is used in new construction in industrial flooring, foundation grouting, roads etc. It is commonly used where the amount of material is not large and where rapid curing can be achieved. The rate of curing epoxy mortar is directly dependent on the ambient temperature. In cold weather, small patches of mortar can be easily heated to heal faster. Clean, dry, specially graded silica sand is used as a filler with a resin-hardener mixture. Sand addition helps to reduce shrinkage, improves friction, thermal shock resistance and low thermal coefficient of expansion make it closer to concrete. Optimum resin and sand ratio to maximum strength is 1 : 7. Epoxy based repair mortar/concrete may not be suitable at the exposed location at high temperatures or where the temperature variation range is large. Excellent mechanical properties and bonds with almost all building materials.

3. Special Mortar :

The following different types of special mortars are available, they are :

(i) Quick setting cement mortar :

- It is basically non-hydrous magnesium phosphate cement consisting of a liquid and a dry powder. Quick-setting cement mortar is made by mixing liquid with dry powder, just like cement and aggregate to make cement concrete.

(ii) Cement-clay mortar :

- Here the clay is introduced as an effective fine-grained additive, which does not exceed 1 : 1 the cement-clay ratio. The addition of soil also improves grain structure, water retaining capacity and workability of mortor and density of mortar. This type of mortar has better casing strength and can be used in thin layers.

(iii) Light and heavy mortar :

- **Light Weight Mortars :** These are prepared from light porous sand and other fine aggregates. They are also prepared by mixing dust wood powder, wood shavings or saw dust with cement mortar or lime mortar. In such mortars, jute coir and hair fibers are cut into pieces of appropriate size, or asbestos fibers may also be used. The bulk density of these mortars is less than 15 kN/m^3.
- **Heavy Weight Mortars :** These are prepared from heavy quartz or other sand. They have a bulk density of 15 kN/m^3 or more. They are used in load bearing capacity.

4. Air-entrance mortar :

- The working properties of the lean cement-sand mortar can be improved by entraining air into it (the air acts as a plasticizer producing minute air bubbles that help with flow characteristics and practicality). Air bubbles increase the amount of binder paste and help fill voids in the sand. The air entraining also makes mortar weight and better heat and sound insulator.

3.4.3 Special Concretes

- These are used for repair and restoration of large damaged areas or for strengthening of an existing structure or to provide some special property to the structure.

Types of Special Concrete :

- Polymer concrete.
- Polymer modified concrete.
- Polymer impregnated concrete.

- Shrinkage compensated concrete.
- Underwater concrete.
- Heat resistance concrete.
- Free flow micro-concrete.
- Epoxy concrete.
- Sulphur concrete.
- Slurry infiltrated concrete.
- Fibre reinforced concrete.

(i) Polymer Concrete : Polymer concrete is a aggregate bond with a polymer binder instead of portland cement because it is conventional concrete. Polymer concrete is typically used to reduce the amount of voids in aggregates. This can be achieved by properly grading and mixing of the material to obtain minimum void volume and maximum density. Entangled aggregates are pre-packed and vibrated in a mold. Mineral voids occur during the curing of Portland cement. Water can be trapped in these voids that can easily attack concrete by freezing. Furthermore alkaline Portland cement is easily attached to chemically aggressive materials, resulting in rapid fixation, using polymers that can cause chemical attacks.

Porosity is caused by air vessels; the water voids due to porosity is due to gel structure. The strength of concrete naturally decreases. It is believed by many researchers that lack of porus increases the strength of concrete. Processes such as vibrations, pressure application spinning etc. have been practiced mainly to reduce porosity. Polymerization is the latest technology to reduce the porosity of concrete to improve the strength and other properties of concrete. The development of solid polymer composite materials for the manufacture of new materials has been directed by combining the ancient technology of cement with the modern technology of polymer chemistry. The materials used in polymer modifying systems are similar to those employed in normal mortar and concreting operations, but for latex/polymer, which is used as a modifier. However these are briefly described as follows :

(A) Cement : OPC is widely used for polymer modified mortar and concrete, including all other Portland cement depending on their applications. However, air entrancing cement is not used as air entrainment occurs due to latex joints.

(B) Aggregates : Aggregates used for general concreting operations are recommended for latex mixes. Aggregates should be clean, sound and have proper grading.

(C) Other materials : Alkali-resistant glass, steel polyamide, polypropylene, polyvinyl alcohol, acramide, and carbon fiber are used as reinforcements. The production of colored latex mortars requires the incorporation of pigments which are alkali resistant and weatherproof. They should not interfere with the stability of latex and the hydration of cement.

(D) Polymers : Polymers are long molecules, made by combining single units called monomers. Polymers are essentially hydrocarbons. The process of conversion of monomers into polymers is called polymerization. In civil engineering, polymers obtained from monomer at ambient temperature are important from a practical point of view.

Polymer concretes are many times more expensive than plane concerts, so they can only be used for special applications.

The **major advantages** of these concerts are :

(i)　Greater failure strain.

(ii)　Good bond with old concrete.

(iii) Better resistance to friction.

(iv) Improved durability and resistance to chemical attack.

Types of Polymer Concrete :

Table 3.1 : Comparative study of mechanical properties of polymer concretes

Sr. No.	Polymer-monomer	Polymer/ aggregate ratio	Density, kg/m³	Compressive strength, N/mm²	Tensile strength, N/mm²	Flexural strength, N/mm²	Modulus of elasticity, GPa
1.	Polyester	1 : 10	2400	117	13	37	32
2.	Polyester	1 : 9	2330	69	–	17	28
3.	Polyester-styrene	1 : 4	–	82	–	–	–
4.	Epoxy + 40 percent dibutyl phthalate	1 : 1*	1650	50	130	–	2
5.	Epoxy + Polyaminoamide	1 : 9	2280	65	–	23	32
6.	Epoxy-polyamide	1 : 9	2000	95	–	33	–
7.	Epoxy-furan	1 : 1*	1700	65	7	0.1	–
8.	NMA-TMPTMA	1 : 15	2400	137	10	22	35

Table 3.2 : Comparison of mechanical properties

Properties	Unit	Cement concrete	Polymer cement concrete	Epoxy concrete	Epoxy cement concrete
Compressive strength	kg/cm²	200 - 250	250 - 300	800 - 1000	500 - 570
Tensile strength	kg/cm²	20 - 25	30 - 35	100 - 130	110 - 120
Flexural strength	kg/cm²	50	60	200 - 300	100
Density	kg/m³	2400	2200	2000	2350
Modulus of elasticity	kg/cm²	$3 - 4 \times 10^5$		1.5×10^5	2.5×10^5
Coefficient of linear thermal expansion	cm/cm°C	$10 - 15 \times 10^{-6}$	–	$25 - 40 \times 10^{-6}$	$14 - 16 \times 10^{-6}$
Band strength	kg/cm²	Less than its cohesive strength	10 - 20	Min. 30 thereafter concrete failure	10 - 20
Abrasion resistance	Taber abraser mg/cycle	4.19	2.70	Less than 0.15 – 0.30	2.50

(ii) Polymer impregnated concrete (PIC) : PIC is one of the widely used polymer composite. Types of monomer used are Methyl-metha-crylate Styreme, acrylo nitrate, T-butyl styrene Other thermo plastic monomer. The amount of monomer that can be loaded into a concrete specimen is limited by the amount of water and air that has occupied the total voids space. The monomer can be loaded in the concrete by vacuum or thermal drying, the later being more practicable for water removable because of its rapidity. Before the specimen is soaked in monomer, the application of the pressure can also been used to reduce monomer loading time.

Table 3.3 : Comparative study of mechanical properties of PIC

Sr. No.	Monomer	Polymer dose, % of mass	Strength, N/mm^2			Modulus of elasticity, GPa
			Compressive	Tensile	Flexural	
1.	Unimpregnated	0	35	2	4	19
2.	MMA	4.6 – 6.7	142	11	18	44
3.	MMA + 10 percent TMPTMA	5.5 – 7.6	151	11	15	43
4.	Styrene	4.2 – 6.0	99	8	16	44
5.	Chlorostyrene	4.9 – 6.9	113	8	17	39
6.	19 percent polyester + 90 percent styrene	6.3 – 7.4	144	11	23	46
7.	t-butyl styrene	5.3 – 6.0	127	10	–	45
8.	60 percent styrene + 40 percent MPTMA	5.9 – 7.3	120	6	–	44
9.	Acrylonitrile	3.2 – 6.0	99	7	10	41
10.	Vinyl chloride	3.0 – 5.0	72	5	–	29
11.	Vinlidene chloride	1.5 – 2.8	47	3	–	21

(iii) Partially impregnated and surface coated polymer concrete : The admixing of polymer latex in cement mixtures modifies the following physical and mechanical properties :

(a) **Workability :** Generally, polymer modified mortar/polymer modified concrete (PMM/PMC) has better workability than conventional mortar/concrete.

(b) **Water retention :** PMM/PMC has improved water retention property significantly over ordinary mortar/concrete. The need for water-curing is greatly reduced and it needs to be specified to suit the polymer type, its proportion, and the method of curing.

(c) **Bleeding and segregation :** A better resistance to bleeding and segregation even if they have better flow capacity.

(d) **Increased resistance to crack propagation :** Micro cracks readily occur in ordinary stressed hardened cement paste. This leads to poor tensile strength and fracture toughness. Whereas, in latex-modified mortar and concrete, it appears that micro cracks are bridged by polymer films or membranes, which prevent crack propagation as well as a strong cement hydrate – aggregate bond develops.

(e) **Strength :** PMM/PMC with styrene butadiene polymer (SBR) latex has a noticeable increase in tensile and flexural strength, but hardly any improvement in its compressive strength compared to ordinary mortar/concrete. An increase in polymer content or polymer – cement ratio (defined as the weight ratio of the amount of total solids in polymer latex to the amount of cement in a latex-modified mortar or concrete) leads to an increase in flexural tensile strength and brittleness. However, the monolithic network formed due to excess air penetration and polymer inclusions causes discontinuity of the structure, which reduces its strength.

(f) **Chemical resistance of PMM/PMC :** It depends on the polymer, polymer cement ratio and type of chemicals. Most PMMs and PMCs with styrene butadiene polymers (including modified acrylic etc.), are attacked by strong organic and inorganic acids and sulfates but they resist alkalis and salts other than sulfates. Their resistance to chloride, fat, and oil is also considered good, while they have poor resistance to organic solvents.

(g) **Temperature effect :** The strength of PMM/PMC depends on the temperature. They usually show a rapid decrease with an increase in temperature. Most thermoplastic polymers have a glass transition temperature of 80°C to 100°C.

(h) Shrinkage : Drying shrinkage of PMM/PMC can be larger or smaller than standard mortar or concrete depending on the type of polymer and polymer-cement ratio used. The higher the polymer ratio, the less is the drying shrinkage.

(i) Water proofing quality or permeability : PMM/PMC has a structure in which larger pores are filled by polymer or sealed by continuous polymer flow. The sealing effect and porosity due to polymer films or membranes formed in the structure also provide a substantial increase in water proof or water tightness as well as resistance to chloride ion penetration, moisture transmission, carbonation and oxygen diffusion chemical resistance, and freeze-thaw durability. Such effects are promoted with increasing polymer-cement ratio upto a certain level of polymer loading.

(j) Adhesion or bond strength : A very useful aspect of PMM/PMC is their improved adhesion or bond strength for various sub-strata compared to conventional mortar/concrete.

(k) Abrasion resistance : PMM/PMCs have better abrasion resistance than conventional mortar/concrete.

(l) Durability and non-degradation : These materials are generally bio non-degradable after total polymerization. However, some polymers decompose into any form of energy such as ultra violet rays, heat etc., especially "styrene" based materials undergo such rapid disintegration and degradation and therefore advised to avoid. While "acrylate" based materials are accepted due to their non-degradable and strong properties.

(m) Sulphur concrete : Sulphur infiltration concrete can be employed in precast industries. Sulphur infiltration concrete should be of considerable use in industry situations where high corrosion resistant concrete is required. This method cannot be easily applied to cast – in place concrete sulphur impregnation has shown area improvement in strength.

3.5 JOINT SEALANT

- These are defined as, "liquids that are applied to a surface of hardened concrete to prevent or reduce the penetration of liquid or gaseous media, such as gases, water, aggressive solutions, gases like carbon dioxide etc. during service exposure conditions".

- The patch repair is based on patch work and chemical effects are not anticipated wherever polymer modified cementitious mortar or epoxy mortar is used. However, to protect the remaining areas from environmental attacks and to avoid subsequent repairs in such areas, the application of seal coat becomes necessary.

- Generally, these protective seal coats are appropriately pigmented so that in addition to protection, the aesthetics of the structure are also taken care of simultaneously. Various coatings such as polyurethane, epoxy, almonds, chlorinated rubbers, acrylic emulsions can be used for this purpose.

3.5.1 Points to be Considered at the time of Selection of Joint Sealant

The selection should be made considering the following points :

- Adhesion to the surface.
- Compatibility with surface alkalinity.
- The coating should be sufficiently impermeable but at the same time, it should allow moisture evaporation from the concrete mass.
- Resist various aggressive environmental chemical attacks.
- Expected life of treatment.
- Ability to cover surface irregularities.
- Ease of application and aesthetics.
- It should adjust the degree of movement and movements occurring on the joint by deformation.
- It should be able to adjust the movement for cyclical changes such as temperature, humidity, vibration etc.

- It should have a good impact without cohesion failure, resisting load, compression, tension and impact stress.
- The flow must be resisted due to gravity.
- It should have good flexibility at all service temperatures.
- It should be durable. It should not be adversely affected by aging, weathering, freezing of water, light, water vapor, growth of fungi and human damage.
- It should have good fire resistance and fuel resistance.

3.5.2 Functions of Sealants

- Accommodate continuing changes in the size of a joint.
- Accommodate movement of various elements.
- Exclude rein, snow and wind from the interior of a building.
- Resist freeze/thaw influences.
- Resist abrasion and water flow through the joint.
- Exclusion or retention of water or liquid.
- Prevent the ingress of anything which would interfere with movement in horizontal joints (e.g. in floors, debris) or cause damage to the building components.
- Exclude chemical or biological contaminants from the joint to preserve the background or interior from chemical attack, or to maintain hygienic conditions.
- Provide sound insulation.
- Prevent heat losses arising from movement of air through the joint.

3.5.3 Type of Sealants

- Oleo resinous mastic sealant.
- Bitumen and rubber/bitumen-based sealant.
- Acrylic resin sealant : Solvent base and water base.
- Flexible epoxide sealants.
- Polysulphide sealants.
- Polyurethane sealants.
- Silicone sealants.
- Hot-pouredd sealants.

(i) **Oleo Resinous Mastics :** These are materials which form the surface skin after application, thus protecting the main body of material underneath. These are commonly referred to as mastics. Both oil-based and butyl-rubber reinforced materials are available. Oleo-resinous mastics are suitable for joints where very little movement is expected. These materials have a limited life, but this can be extended by regular painting. They have good adhesion. No chemical bonding takes place with the substrate. Typical applications include window perimeter pointing, but they should not be used on microporous finished timber or PVC. The buytl-rubber reinforced materials are commonly used for sealing lap joints in metallic building panels.

(ii) **Bitumen and Rubber/Bitumen-based Sealants :** Bitumen-based sealants are thermoplastic, and they retain a degree of flexibility. Enhanced elasticity and flexibility can be obtained by modifying the bitumen with rubber. Rubber/ bitumen sealants are essentially plasto-elastic in nature and can develop a non-tacky surface within 24 to 48 hours after application. Bitumen and rubber modified products are used in roofing, water-retaining structures and areas where compatibility with other bituminous materials is desirable.

(iii) **Acrylic Resin Sealants :** Basically, types of acrylic resin sealants are in common usage : solvent and water based. Solvent based acrylic sealants are primarily used for refurbishment work where their tenacious adhesion to surface which are difficult to clean is a distance advantage. They are thermoplastic materials with plasto-elastic properties. They are used externally in the vast majority of cases. Appearance and durability can be improved by painting. Water-or emulsion based acrylic materials are widely employed as sealant for painting of window and door perimeters particularly internally. If used externally, care should be taken to avoid early contact with water or rain, because this can lead to washing of the sealant.

(iv) Flexible Epoxide Sealants : These are based on epoxy resins, and varying degrees of flexibility can be achieved by the addition of other polymers or extenders. Epoxide sealants are normally available as multi-component products which, when mixed, get cured at ambient temperature. Epoxide sealants are predominantly plastic in behavior. Although they show a degree of flexibility at room temperature and above, they become rigid at low temperatures.

(v) Polysulphide Sealants : These are available in one and two component versions. The single part materials cure on exposure to atmospheric moisture. They are essentially elasto-plastic in nature. The polysulphide sealants require site mixing and they do not rely on atmospheric moisture for curing. Curing is taken place uniformly throughout the body of the material. Polysulphide sealants have been widely used in normal building and civil engineering applications over many years.

(vi) Polyurethane Sealants : Polyurethane sealants are also available in gunapplied and flow grades as one or two component systems. These can exhibit a wide range of properties and normally their cure rate is rapid. When cured they have elastic properties. Two component polyurethance sealants are chemically curing products not relying on atmospheric moisture to effect curing. Some polyurethane sealants are tar or pitch modified. Polyurethane sealants are used both in building and civil engineering applications.

(vii) Silicone Sealants : These sealants are commonly available as single component materials with a variety of cure systems. Curing takes place by reaction with atmospheric moisture. Skin formation is generally rapid and the acid-curing version cure quickly in depth. Recent advances have produced a wide variety of cure systems. Specific information regarding these materials and their intended applications should be obtained from the relevant sealant manufactures. Silicones are versatile enough to produce a range of properties with respect to modulus and elongation. Applications include sealing of curtain wall panels, pointing of PVC components, glazing and sealing of joints in road pavements.

(viii) Hot Poured Sealants : These comprise of bitumen, rubber with bitumen and pitch with polymer combinations. These are primarily used in horizontal joints in road pavements, water retaining structures, and subways. It is essential that the correct equipment should be used to apply these materials. Temperature of the sealant is a critical parameter, and needs to be accurately controlled.

3.5.4 For Effective Sealant Field Performance, the Following Points are Very Important

- Selection of correct sealant.
- Calculate the correct width to depth ratio.
- Proper cleaning of the joint faces.
- Through details on joints.
- Appropriate primer based on surface.
- Use of non-impermeable backup material.
- Provision of bond breakers.
- Appropriate application tools and techniques.
- Qualified applicants.
- Adequate field inspection.

3.6 GROUTS

- A grout is "a fluid material which is designed to be introduced into the cavity for the purpose of filling and subsequently hardening to give specific physical properties". The process of grouting is to fill voids, gaps or cavities with material having certain specific properties. Grouts are most common materials for repair of building elements.

3.6.1 Properties of Grouts

- The materials of grout should not shrink, but should expand fill gaps.
- Grout must have ease of filling with enough fluidity.
- Grout must have capacity to bear loads by having compressive and tensile strength and vibratory load resistance.

3.6.2 Types of Grout

- Cement grout.
- Cement sand grout.
- Cement sand grout with additive.
- Polymer modified cement grout.
- Epoxy/Polyester Resins.
- Mastic.

(a) Cement Grouts :

- Cracks wider than 1 mm can be sealed by brushing in dry cement followed by spraying lightly with water. For cracks wider than 2 mm, it is preferable to use cement water slurry grout. Alternatively, cracks can be chased out to a width of 5-10 mm and pointed with cement and sand mortar.

(b) Cement-sand with Additive Grouts :

Cement-based grout is made of mixed water and cement, which is sometimes also added with sand and admixtures. It is commonly used for soil improvement such as dam curtain walls using jet grouting methods, for repairing the damages on concrete and masonry, or for the construction of preplaced aggregate concrete. Currently, there are hardly any available practical methods which can be used to carry out the grout mix design.

(c) Polymer Modified Cement Grout :

- Low viscosity liquid polymers can be used in a similar way as cement grout for repair of cracks. Cracks may be brush application or by temporary ponding with liquid polymers. When no further materials will penetrate the crack, the surplus materials are removed by wiping.

(d) Epoxy Resin Injection of Cracks Under Pressure :

- Cracks in reinforced concrete greater than 0.3 mm may require sealing by resin injection to prevent ingress of moisture, oxygen and harmful agents. Before deciding upon the most appropriate method and material for repairing the crack, it is imperative to establish the nature and cause of the cracking. Cracking is caused normally by tensile stresses and if these stresses re-occur after the crack repair, the concrete may crack again.
- Low viscosity epoxy resin system with a viscosity below 6 stokes at 20°C is generally used for repair of cracks. Low viscosity acrylic or polyester resins are found to give lower bond strength, especially under damp conditions. The pressure of injection depends upon the width and depth of crack.
- When repairing cracks of 2 mm or more it is not always possible to seal the outlets. In such cases, low viscosity resins are not preferred and thixotropic resins are used. These modified resins flow readily into relatively fine cracks under low pressures, but stop flowing immediately when the pressure is released.

(e) Repair of live Cracks with Mastics :

- Normal grout injection will not be effective when repairing live crack, as the crack further widens or new crack opens at the interface of existing hardened grout and concrete. In such situations, materials that can accommodate considerable strains against cracking are to be selected.
- When the anticipated future movement of the crack is small (less than 15% of its width) apply 'mastics'. Viscous fluids such as non-drying oils or low melting soft asphalts along with fillers/fibres, to the cracks. When the movement is more (greater than 25%) apply 'thermoplastics' such as asphalt, coal tar or pitches. However, these have limited durability under ultra-violet light rays. Elastomers have advantage over mastics and thermoplastics and are more commonly used in recent times. Elastomers include polysulphides, polyurethanes, certain silicones and acrylics.

3.7 WATERPROOFING MATERIAL

- Water is an important resource, but also the biggest enemy of building materials, since rain, groundwater and surface water can cause rapid and extensive damage to buildings. The solution is contained in water-repellent building materials with sealing properties such as cementitious slurries modified with dispersible polymer powder.

- Water, in the form of liquid or vapor, is the most destructive weathering element for buildings constructed from materials such as concrete, masonry and natural stone. Waterproofing techniques must therefore maintain the integrity, functionality, and utility of a structure for its entire life.

- Due to the harsh monsoon condition, there is special challenge for waterproofing system in India. To eliminate all possible causes of water intrusion, the exterior walls, roof and basement of a building should be completely covered with waterproof material. All waterproofing measures must be part of an entire system and must be fully negotiated to be fully effective in preventing water ingress.

- Should one component of the system fail or not fully interact with all other parts, leakage may occur. Potential damage, degradation, and unnecessary repairs to building construction can be avoided, by controlling groundwater, rainwater and surface water, as well as transporting humidity in the form of water vapor.

- Traditional sealing and waterproofing systems include bituminous materials, plastic waterproof foil, and metal tape for internal and external applications. In addition to these systems, products based on reactive resins, purely dispersion-bound, pasty products and cementitious waterproofing membranes are now widely used to seal and protect the exterior surfaces of buildings and structural components against the action of water and moisture.

3.7.1 Water Proofing Materials for Roof

    (a) Polyisobutylene (PIB) Sheeting.

    (b) Glass Fibre Reinforced Plastics (GRP).

    (c) Bitumen and Bituminous Emulsions.

    (d) Latex Cement Coatings.

    (e) Epoxy and Polyurethane (PU) Resin Coatings.

    (f) Acrylic Polymer Modified Cementitious Coating.

    (g) Polymeric Modified Bituminous Membrances.

    (h) Plymer Based waterproof Coating.

    (i) Butyl Rubber (BR) Sheeting.

(a) Polyisobutylene (PIB) Sheeting : It is black, flexible waterproofing sheeting similar to the BR sheet with improved properties. It has a low restitution as a result of which when it is stretched, there is lesser tendency to regain its original shape and residual stresses are reduced. The PIB sheets have increased life in ultraviolet light, ozone and a wide range of chemical environment. PIB sheets do not support fungi growth. These sheets have longer abrasion resistance compared to BR sheets and are not recommended for waterproofing of heavy foot traffic roofs. The sheets are available in 0.6 to 2.0 mm thickness. For general purpose 0.8 to 1.0 mm thickness is adequate. The joints are solvent welded. The solvent softens and activates the sheets at the joint and completely evaporates, leaving strong and durable joint.

(b) Glass Fibre Reinforced Plastics (GRP) : It is a composite material of polyester resin and glass fibre, usually referred to as GRP. GRP is applied on liquid retaining structure to prevent leakage and to protect against chemical attack. Usually GRP is spray applied, with the resin, hardner and glass fibre through a 3-point nozzle gun. Special technique is employed to ensure that the fibres are completely covered by the resin.

(c) Bitumen and Bituminous Emulsions : Bitumen is frequently used in protection of leaking or damaged concrete roof slab against ingress of water and water borne aggressive agents. Minimum 2 coats are to be applied. The second coat at right angles to the first helps to eliminate pinholes. It is cost effective and easy to apply for drinking water retaining structures, special grade bitumen which is non-toxic and taint free is used.

(d) Latex Cement Coatings : For water retaining structure of low porosity the inside can be treated with a coat of 2 parts of cement and 1 part latex by weight. It gives a coverage of 0.8 m^2/litre of the paste. The cleaned concrete surface is given a brush application of the latex – cement rendering. On drying, second coat is applied at right angle. It is simple to use. It is relatively cost effective.

(e) Epoxy and Polyurethane (PU) Resin Coatings : These are basically two or more component organic polymers and can be formulated to give coatings of wide range of characteristics and uses, such as water-proofing chemical resistance bond, viscosity pot life curing and coloring. Concrete surface should be free of dust and oil to ensure bond with repair a material. For water proofing, a minimum thickness of 0.5 mm in coats is recommended. It can be applied by brush or spray.

(f) Acrylic Polymer Modified Cementitious Coating : These are designed to resurface and even out variations in concrete surface. These provides a long lasting barrier to water borne corrosive salts and atmospheric gases providing resistance to concrete decay. These provide a seamless waterproofing coating suitable for use in water tanks, reservoirs, fountains, roofs and sealing the bar holes to ensure water tightness. These effectively seal concrete masonry walls, bridges and static shrinkage cracks. These provide a tough and durable coating which cannot be easily damaged or worn away. These have excellent weather resistance properties and are suitable for exterior applications.

(g) Polymeric Modified Bituminous Membranes : These membranes consist a centre core of High Density Polyethylene film (90 micron) which is the heart of the membrane which forms complete barrier against water and moisture. The center core is protected on both sides with a high quality polymeric asphaltic mix with properties of high penetration high heat resistance and high bonding strength with substrate and in overlaps. The polymeric asphaltic coating film is protected on bottom side with thermofusible High Molecular High Density Poly-Ethlene (HMHDPE). This aids the thermo fusing process and also reinforce the non-permeable quality of the entire membrane including overlaps and joints. The membrane is finished on top with Nitrocellulose Lacquered Embossed Foil of 75 micron thickness. The resultant membrane has elongation exceeding 200% to absorb any structural movement. It also has high tensile strength and is extremely flexible and pliable to adopt to contours. It is lightweight and is ideal for precast structures. It is also highly sound absorbent on top of light CGI roofing. It is available in rolls of 1 m width and 15-20 m length.

(h) Polymer Based Waterproof Coating : These are single component based on modified polymers and are free from tar bitumen and solvents. The product is available in paste like consistency. The product is applied on cementitious substrates by means of brushed or roller after one coat of primer. The dried coat forms a seamless membrane which is flexible and elastic in nature having a breaking elongation of 82%. The coating has the breathing properties. The ciating has adequate resistance to U/V rays and does not crack or flake after multiple thermal cycles.

(i) Butyl Rubber (BR) Sheeting : It is used for water proofing of roof slabs and treating leaking roofs of buildings. It is tough, black and flexible sheeting with considerable abrasion resistance. It requires careful laying as it can be It requires careful laying as it can be punctured by sharp edges of concrete surface while laying. The sheets are of thickness 0.5 mm to 2.0 mm. These have good abrasion resistance and durability.

3.7.2 Steps in Waterproofing

- The application of this system broadly includes the following steps :

Basic Steps for Preparation :

    (i) Roof surface preparation : The surface shall be cleaned to remove all dust, foreign matters, lose material or any other deposits of contamination. The cracks and depressions will be filled by filler (cement sand mortar as per supplier's recommendation). Prepared surface shall be thoroughly prewetted for one hour.

(ii) Preparation of polymer modified cementitious slurry : Dry mixture and liquid mixture shall be mixed in desired proportions as per the recommendation of the supplier. The mixture will be stirred well, until there are no bubbles in the mixture. Any lumps found in the mixture will be removed.

(iii) The first coat of polymer modified cementitious slurry shall be applied on a wet clean surface by brush.

(iv) After this, the fiber glass cloth will be placed on the first coat of polymer modified cementitious solution.

(v) The second coat of polymer modified cementitious slurry will be placed on fiber glass cloth.

(vi) Polymer modified cementitious brush topping will be applied to the second coat of polymer modified coating.

(vii) On brush topping, screed concrete,1 : 2 : 4 admixed with suitable integral waterproofing compound 25 to 40mm thick to a slope of 1 in 100, aggregate size below 10 mm shall be maintained with a maximum water cement ratio of 0.45. The above system may differ slightly from case to case depending on the instruction of the supplier of the water proofing system. There is no relevant Indian Standard/Other Code of Practice for this system. Therefore, work should be done as per the instructions of the manufacturers/suppliers. Users are advised to collect the complete literature from the manufacturer and study it carefully before application of the treatment.

Steps of Application :

(viii) Provision of slope and cleanliness of roof surface : Before the application of water proofing treatment, the roof surface should be provided with a minimum slope of 1 in 100 with plain cement concrete. After the provision of slope, all the preparation work such as filling cracks with cement sand slurry, provision of sufficient number of drain outlets, provision of 75 mm fillet at the junction of roof slab with parapet wall, provision of groove/chase in parapet wall etc. described in IS : 3067 must be completed.

(ix) Laying a coat of cold applied bitumen primer @ 0.2 to 0.4 l/sq. m. on the entire roof surface.

(x) Laying 85/25 grade hot blown bitumen @ 1.2 kg/sq. m. all over surface.

(xi) Laying 2.5-3 mm thick polymer modified bituminous membrane with non-woven polyester fibre glass mat reinforcement, applied by torch with sealing all the joints.

(xii) Laying 85/25 grade hot blown bitumen @ 1.2 kg/sq. m. all over the surface.

(xiii) Final finish with china mosaic tiles on a 15 mm thick grey cement plaster bed.

The method of application may vary slightly depending on the product and manufacturers' recommendations. Since there is no relevant IS Code of Practice for this water proofing system, work should be done as per the recommendations of the manufacturers. Users are advised to collect all the literature from the manufacturers and study it thoroughly before applying.

3.7.3 Field Requirements for Acrylic Waterproofing

- These systems must block the bulk water movement in the substrate they protect, so the coating product itself plus the method of application must ensure a watertight membrane and because they apply to a wide variety of substrates, including asphalt-based roofs, plastic and rubber single ply roofing, sprayed polyurethane foam, metal, and concrete, the chosen coating is designed to improve its waterproofing and safety performance on a specific substrate type.

- To ensure this long-term performance, a coating system has to withstand weathering well, which resists damage from the sun's UV and hot infrared radiation. Coatings must have inherent properties to handle construction movement and temperature changes that cause expansion and contraction stresses. Beyond these factors, foot traffic on terraces, regional hail activity, and snow and ice buildup require additional strength and flexibility.

- Over the years, the technology around acrylic coatings has matured. Raw material manufacturers and top coating formulators understand the key components of proven coatings and each component plays a role in meeting the above performance requirements to ensure long-lasting waterproofing. Below are several specific coating components and their contribution to sustainable performance based on years of exposure on actual building.

- **Acrylic Polymer Resins :** Acrylic polymer is the major determinant of important coating properties such as water resistance, impact and tear strength, ductility, a given substrate, and overall durability, which determines the quality of the acrylic polymer vis-à-vis. These performance parameters are the chemical composition of the polymer coupled with the actual synthesis process by which the raw materials (monomers) are reacted to produce the final emulsion polymer. 100% acrylic compositions are preferred as co-polymers such as styrene acrylic and vinyl acrylic. Styrene acrylic often shows inferior weathering and colour instability due to UV attack of aromatic chemistry in styrene; Vinyl acrylic, polymers often used in cheaper interior paints, show poor water resistance due to the hydrophilic properties of vinyl chemistry. 100% acrylic polymers, however, are completely transparent to UV rays, so they act as durable binders that can hold the coating film together for long slopes. 100% acrylic has proven superior adhesion properties, being the major components of adhesion, architectural sealant, and tape and label adhesives and properly engineered acrylic polymers provide the right balance between strength and stretch, also known as toughness. For instance, typical paints are not waterproof coatings and always develop small cracks over a time due to building movement and temperature fluctuations. The paints are based on "hard" acrylic polymers. However, waterproofing coatings must be formulated with so-called "elastomeric" acrylic polymers, which provide low-temperature flexibility while retaining other critical strength, adhesion and waterproofing properties. The bottom line is that the incorporation of suitable acrylic polymer is an essential determinant in the overall performance of an elastomeric acrylic coating.

3.8 SURFACE COATINGS FOR CONCRETE PROTECTION

- Surface coatings are applied to concrete surface to prevent ingress of moisture and provide protection against aggressive environment. Preparation of the substrate prior to the use of surface coatings is necessary. All concrete surfaces whether new or old will have weak cement laitance, oil, grease, dust and other contaminations. If these are not removed, the coatings may fail in bond.
- The functions of surface are of two types :
  (i) To act as a barrier against aggressive environmental attack.
  (ii) To improve the visual appearance in addition to giving the surface self cleaning properties.
- Surface treatment can be used for :
  - Enhancing appearance (colour, texture, opacity, cleanability and reflectance).
  - Improving chemical resistance to sulphates, acids, brewery and dairy products.
  - Controlling ingress of chlorides, oxygen, carbon-dioxide, water vapour and moisture.
  - Improving mechanical and physical endurance such as resistance to abrasion, impact and skidding.
- Surface treatments are expected to perform satisfactorily under widely variable and extreme service conditions such as hot and humid, cold wet or dry cycles, covered and underwater. They often have to fulfill several distinct requirements at the same time. Depending upon the condition of the concrete, requirements of the structure and the environment, the factors to be considered for concrete coating selection are as :
  - Resistance to water ingress.
  - Resistance to carbon dioxide diffusion resistance.
  - Resistance to chloride ingress.
  - Water vapor diffusion resistance.
  - Ultra violet light resistance.
  - Elastic/crack bridging abilities.
  - Chemical resistance.
  - Abrasion resistance.

- Ease of application.
- Service life required.
- Ease of overcoating.
- Aesthetic appearance.

- The coating of elements normally includes floors, bured pipes, exteriors of structures and tanks, tank interiors, marine structures, bridge piers and superstructure and nuclear power contaminant vessels. These structures need protection against a variety of conditions such as food chemicals, fats, oils, corrosive soils, chemical plant wastes and fumes, fungus and even bacteria.

Common coatings used for these purpose are :

(i) Bituminous Cutbacks : These coatings are solvent solutions of coal tar or asphalt. The coal tar cutbacks have better chemical resistance and better water impermeability than the asphalts. On the other hand, the asphalt cutbacks have better weather resistance and sunlight resistance. A thin coat is generally applied as primer, followed by heavier coats over the surface to protect water penetration. Some bituminous cutbacks are used as concrete penetrates. These are applied as thin material coat and allowed to soak into the concrete, increasing the surface density and thereby minimizing water penetration and concrete spalling. Bituminous coatings are also available as water emulsions which cure by water evaporation and resin coalescence.

(ii) Chlorinated Rubber Coating : Chlorinated rubber coatings have water, chemical and alkali resistance. They are lacquer type and hence dry rapidly to form a good resistant film. They have very good adhesion to concrete. They are extensively used as coating for concrete water tanks, swimming pools and are specially formulated for tough, abrasion-resistant concrete floor enamels. They perform well under high humidity conditions and can be easily recoated. Chlorinated rubber coatings are vulnerable to rodent attacks, vegetable oils/ greases and hence these coatings are not recommended for sewer or linings.

(iii) Vinyl Coatings : Vinyl coatings have been used for many years as a coating for different concrete structures, from tank lining to nuclear power plant installations. They have good chemical resistance to both acids and alkalies. In view of their high molecular weight, their application needs to be preceded by a thin primer for maximum penetration into concrete surfaces. Vinyl coatings are lacquer type which dry very fast, hence it is advisable to coat them in cooler temperatures. Vinyl coatings generally have relatively low solid content so multiple coasts are needed to build required thickness. For exterior applications, vinyl acrylic top coatings give better weather resistance.

(iv) Epoxy Coatings : Epoxy coatings are formulated with liquid epoxy resin, liquid curing agent and highly penetrating solvent for good performance. These coatings allow the epoxy resin to penetrate deeply into the concrete surface, react to increase the density as well as the strength of the concrete surface providing excellent adhesion and chemical resistance.

- Epoxy coatings are extensively used for a variety of structures such as tank lining, floor, pump base and nuclear fuel storage areas etc. There are several types of epoxy formulations used in coating concrete surfaces. These could be grouped into the following three types.

  1. Thin epoxy coating which can be applied over sand-blasted concrete surface. These are usually solvent based epoxies with relatively high molecular weight resins. They are similar to many epoxies used for steel coating.

  2. Thin thixotropic liquid epoxy-based primers can be applied to sand blasted surface. They provide both, a good surface and a base for other epoxy topcoat.

  3. Thick epoxy coating can be applied by trowel or by spray. This can be applied directly to a clean unprepared concrete surface. This fills the concrete surface imperfections and can be used alone or with additional coats of conventional epoxy topcoats.

- Epoxy coatings can be cured by either amines or by polyamides and should be applied over smooth concrete which has been etched or sandblasted to a fine sand paper finish. The thickness of the coating usually ranges between 200 to 375 microns.

(v) Coal Tar Epoxies : Coal tar epoxies can be classified as the fourth type of epoxy coating. They have both properties of coal tar and epoxy. They have good adhesion to concrete and better chemical and abrasion resistance. These are easy to apply either by conventional roller or by airless spray. These are normally applied in two coats, having coating thickness of 375-500 microns. Amine-cured coal tar epoxied can effectively withstand the action of severe corrosion, bacterial, H_2S and acid condensate.

(vi) Integrated Four Coat System : Central Electro Chemical Research Institute (CECRI) has developed an integral four coat system for coating of concrete structures consisting of epoxy polyamide ion oxide primer, epoxy polyamide MiO undercoat, epoxy polyamide TiO_2 top coat and aliphatic polyurethane sealer coat to prevent corrosive attack from aggressive marine and industrial environment. Epoxy-based coatings are known for their ability to penetrate deeply into the concrete surface, thereby increasing the strength of the concrete surface. They are also well known for their high alkali resistance and have good adhesion and compatibility with other epoxy coatings as well as polyurethane coatings. Water absorption studies carried out on bare concrete surface, general painted concrete structure and CECRI formulation indicated that the CECRI coat system has the minimum water uptake value and minimum permeation to ions.

3.8.1 List of Various Types of Coatings and Their Performance Characteristics

Table 3.4

Coating Type	Characteristics
Epoxy coatings and modified epoxy coating such as coal tar epoxy, epoxy-phenolic (IPN etc.)	Excellent adhesion to concrete, abrasion impact resistant. Impermeable to organic and inorganic chemicals, water and chloride ions. Excellent resistant to corrosion and steel reinforcement. Generally used for internal applications.
Chlorinated rubber coatings	Resistant to heat, sunlight and weather, moderately resistant to acids, alkalis. Not recommended for immersed conditions. Adhesion to concrete is good.
Acrylic coatings	Higher permeability, a life of three years may be possible.
Polyurethane coatings	Excellent UV resistances abrasion and cracking resistance, has high elasticity and resistance to biological department.
Bitumen or Tar products.	Provide excellent protection; however aesthetics is affected.
Products based on cement sand and asphalt	Showed satisfactory performance for more than six years in tropical marine atmosphere at extreme exposure condition.

3.9 ADDITIONAL REPAIRING MATERIALS

1. **Polyester putty or 1 : 3 cement sand mortar :**

- A polyester-based putty compound is used for rapid bedding, bonding and joining of concrete, brickwork and masonry, rapid joining of precast concrete pipes and units and rapid fixing of steel inserts in precast concrete units.

- Polyester putty is a two-component, rapidly setting compound available in summer and winter grades, which rapidly gains the strength to exceeding parental concrete within an hour. Bonded units can be returned to service within one to two hours.

- The product's mix design allows for full pack mixing or part mixing when required for small jobs. Mixed putty has good adhesion to most construction materials and is resistant to a wide range of chemicals including petroleum products plus road and aircraft runway de-icing salts.

2. **Plastic/Aluminum nipples :**

- Packers/nipples must be in the grout used for the injection of epoxy and polyurethane foam into a pipe inserted in pressures greater than 20 kg/cm (maximum pressure 500 kg/cm) on concrete structures. We manufacture injection packers in various specifications which are essential in the repair and reinforcement of concrete structures depending on the thickness of the concrete structures and on site situations.

- **Advantages of packer :**
  - It is possible to force a large amount of resin at high speed, which ensures through injection.
  - Because backwater check valves are placed on areas of the bolt, it is possible, irrespective of hardening of resin, to snap the severed parts of the nuts immediately after the injection work is completed.

Product Description :

PVC cement slurry injection packers :

Size : 100 mm insertable

Total length : 130 mm

Diameter : 10 mm or 3/8" with knob to stop cement inside from coming out

Wall thickness : 0.5 mm

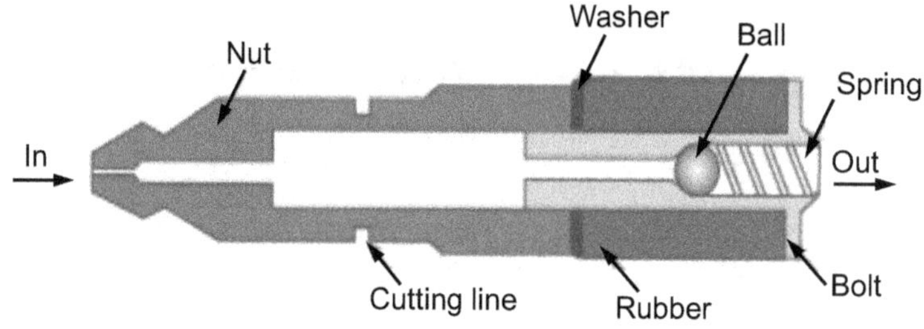

Fig. 3.1

Aluminum grouting nipples :

Length (mm)	80	85	100	150
Diameter (mm)	10	13	13	13
Drill Diameter (mm)	10	13	13	13
Net Weight (g)	12.7	22.5	24.5	30.0
Gross Weight (g)	13.4	23.0	25.0	31.7
Working Pressure (bar)	100	100	100	100

3. Galvanized steel wire fabrics :

- The principles of choosing architecture wall, steel mesh are summarized based on the usage effect.
- When we need a metal mesh, the first choice is a mesh of welded wire, but not a mesh of woven wire.
- When we need a mesh of electro galvanized wire, the premier electro galvanized steel wire is used.
- When we need expanded metal, the preferred option is cold rolled galvanized expanded metal.
- We choose galvanized steel wire mesh which is galvanized after welded.

A. Options for exterior wall

- When the safety mesh for the outer wall is anti-peeling and anti-cracking. To reinforce the anti-cracking effect, welded wire mesh can be used with anti-cracking fiber mortar. Expanded metal is the first choice for anti-cracking exterior wall. When aiming for the outer wall is to protect, we usually choose butt welded wire mesh. Redrawing welded wire mesh and expanded metal are for connecting the part between the outer wall and the main structure. Electro galvanized welded wire mesh and butt welded wire mesh are for high building exterior wall anti-cracking and extended wire mesh is not intended for high building. The specification depends on the exterior wall density.

B. Metal mesh options for interior walls

- Parts between the column and beam require anti-cracking, and the parts on door holes and buried pipeline, need welded wire mesh. The diameter of the wire is 1 mm - 1.2 mm, and the mesh opening is 15 mm - 20 mm. Galvanized expanded metal is also required. The thickness of both the panel and the strand is 0.6 mm. The mesh opening is 10 mm $\times$ 20 mm. Weight is 0.56 kg/m^2. (For Fig. Refer Colour Photographs).

Table 3.5

Steel mesh types	Material standard	Wire species	Shapes	Specifications (mm)	Using ranges
Welded wire mesh	Carbon steel GB/T 700	Electro galvanized low carbon steel wire	Rolled	Wire diameter : 0.4 – 0.9 Mesh opening : 9.5 – 19	Interior wall
			Rolled	Wire diameter : 2.2 – 4.0 Mesh opening : 25 – 50	Exterior wall
		Electro galvanized low carbon redrawn wire	Rolled	Wire diameter : 0.4 – 0.9 Mesh opening : 9.5 – 19	Interior wall
			Rolled	Wire diameter : 2.2 – 4.0 Mesh opening : 25 – 50	Exterior wall
Butt welded wire mesh	Carbon steel GB/T 700	Electro galvanized low carbon steel wire	Panels	Wire diameter : 2.2 – 4.0 Mesh opening : 25 – 200	Exterior wall
		Cold-drawn low carbon steel wire	Panels	Wire diameter : 2.2 – 4.0 mesh opening : 25 – 300	Exterior wall
Expanded metal	Cold rolled expanded metal GB/T 708	Diamond shape mesh opening, rectangle strand	Rolled	Sheet thickness : 0.4 – 1.0 Strand thickness : 0.4 – 1.0 Mesh opening : $12 \times 25 – 8 \times 16$	Interior wall and Exterior wall
Galvanized welded wire mesh	Carbon steel GB/T 700	Hot-dip galvanized steel wire	Rolled	Wire diameter : 0.9 – 2.2 Mesh opening : $12.7 \times 12.7 – 50.8 \times 50.8$	Exterior wall thermal insulation

4. **Clamping rods :** Clamping rods may be used to support the masonry walls. (For Fig. Refer Colour Photographs).

5. **Wired nails :**

- Special nails are used to clamp the mesh to the masonry wall. These nails hold the mesh in its position with required amount of tension.

6. Ferro-cement plates :

- The ferro-cement plates gives confinement to the masonry. Thus, the increase in strength of masonry is due to increase in cross-sectional area as well as due to confinement. There is a considerable increase in ductility as well, which is important from seismic consideration.

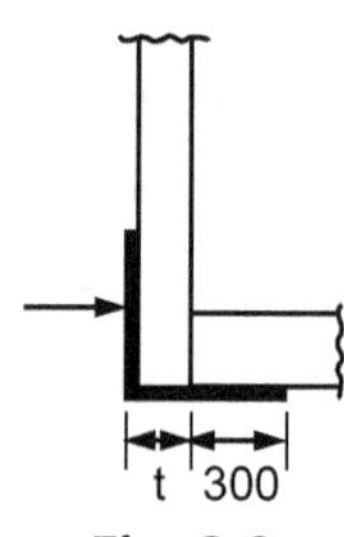

Fig. 3.2

(a) Steel Plates :

(1) The quality of steel plates shall be indicated by their tensile and other strength properties, Young's modulus and other deformation properties, and thermal properties and other material characteristics. The steel plates should be weldable and bonding with adhesives can be ensured when necessary.

(2) The surface of the steel plates should be suitably protected to prevent their quality from changing over time.

(1) SS 400 is a standard material used for steel plates. When SS 400 is used, it must have a JIS standard certificate. When welded joints are used to connect steel plates, the weld should be one for which the required strength can be ensured. When an adhesive is used to bind a steel plate to the surface of the concrete, it will be able to use the bond strength required to secure its adhesive properties. When steel plates of materials other than SS 400 are used, a separate study of the bonding properties and weldability should be done if necessary.

(2) In general, the quality of steel plates will change over time through load action and environmental action. With the bonding method, steel plates are usually placed on the outer surface of reinforced concrete members, so they are affected by changes over time that result from corrosion due to environmental action. As an anticorrosion measure for steel plates, in general, etching primer and zinc-enriched primer are used. Materials whose anticorrosion, adhesion, and other quality requirements have been confirmed should be coated with a suitable coatings thickness.

(b) Continuous Fiber Sheets :

(1) The quality of continuous fiber sheets will be indicated by their tensile strength and other strength properties, Young's modulus and other deformation properties, and thermal properties and other material characteristics.

(2) As a rule, the surface of continuous fiber sheets shall be protected to prevent their quality from changing over time.

(1) Continuous fiber sheets should be checked to confirm that they have strength, elongation capacity, Young's modulus, linear expansion and coefficient of material properties that may depend from an engineering point of view. In addition, the quality of continuous fiber sheets will change depending on the resin with which they are impregnated and bonded, so the material properties of the continuous fiber sheets and fiber-reinforced plastic composite materials should be used for impregnating and bonding.

(2) In general, the quality of continuous fiber sheets will change over time due to load action and environmental action. With the bonding method, a continuous fiber sheet is placed on the outer surface of the structure, so as a rule protective layer of paint, concrete, mortar or the like must be provided to prevent the reinforcing from changing over time. This condition should not be observed, however, if it can be confirmed by the results of appropriate numerical tests and accelerated exposure tests that the quality of continuous fiber sheets does not change over time, even if they are not preserved, or if the decline in quality can be accurately predicted.

7. Fibres :

- All types of fibers have been tried out in cement and concrete. Not all of them can be used effectively and economically. The fibers that can be used are steel fiber, poly propylene, nylon, asbestos, glass and carbon. A fiber is a small piece of reinforcing material possessing certain characteristics.

- Steel fiber is one of the most commonly used fibers, usually round fibers are used. The diameter varies from 0.25 to 0.75 mm. Steel fiber is more likely to rust and reduce its strength. The use of steel fiber leads to significant improvements in the flexibility, impact, and fatigue strength of concrete. It is widely used in a wide variety of structures, especially on the overlay of roads air field pavements, bridges. Thin shells and plates have also been manufactured using steel fibers.

- Polypropylene and nylon fibers are found suitable to increase the impact strength. Asbestos is a mineral fiber and has proved to be the most successful of all fibers because it can be mixed with portland cement. Glass fiber is a recent introduction in making fiber concrete. It has a very high tensile strength of 1020 – 4080 N/mm^2. Alkali resistance shows considerable improvement in fiber reinforced concrete durability. The high tensile strength of carbon fiber poses is 2110–2815 N/mm^2. Carbon fiber will have higher modulus of elasticity and flexibility. The use of carbon fiber for structures such as panels and shells will promise the future.

Aspect ratio :

- Fiber is often described by a convenient parameter called "aspect ratio". The aspect ratio of a fiber is the ratio of its length to diameter. Typical aspect ratios range from 30 to 150.

  ### Types of Fiber :
  1. Steel fibres
  2. Glass fibres
  3. Polypropylene fibres
  4. Asbestos
  5. Carbon

1. Steel Fiber :

- Steel fiber is one of the most commonly used fibers. Round fibers are usually used. The diameter may vary from 0.25 to 0.75 mm. Steel fiber is likely to rust and lose some of its strength. But investigations have shown that, the corrosion of the fibers occurs on the surface itself. The use of steel fiber leads to significant improvements in the flexural, impact and fatigue strength of concrete. It has a very high tensile strength of 1700 N/mm^2. Steel fiber has been incorporated into Shotcrete to improve its crack resistance, ductility and energy absorption and impact resistance characteristics.

  Applications : Industrial flooring, Warehouses, Overlays, Tunneling.

2. Glass Fiber :

- Glass fiber is a recent introduction in making fiber concrete. It has a high tensile strength of 1020 to 4080 N/mm^2. Glass fiber originally used in combination with cement was found to be affected by the alkaline condition of cement. Therefore, alkaline resistant glass fiber has been developed and used by the trade name "CEM-FIL".

  - AR glass fiber has a density that is similar to concrete.
  - It gives better bond between concrete matrix and reinforcement.
  - It prevents cracks.

- Fibers also have elastic modulus that is much higher than concrete. This enables the fiber to provide an effective reinforcement during the hardening phase of the concrete.

  Applications : Noise barriers, Water ducts and channels, Tunnel lining, Railways.

3. Polypropylene Fiber :

- They are good resistance against shrinkage and temperature cracks. It is of low modulus. They have a longer elongation under a given load, which means that they can absorb more energy without fracture.

- Low modulus fibers can be combined with steel fibers which is the latest trend we call hybrid technology. It is applicable to the structure exposed to the atmosphere. These fibers can take care of drying shrinkage while steel cannot perform in wet conditions.

  Applications : Polypropylene fiber can be used for slab on grades, airports, highways, pavements, parking areas, bridge deck overlays, sewer pipes, precast concrete products.

4. Asbestos :

- Asbestos is a mineral fiber and has proved to be the most successful of all fibers because it can be mixed with Portland cement. The tensile strength of asbestos varies between 560 to 980 N/mm^2. The compound product called asbestos cement has significantly higher flexural strength than Portland cement paste. For unimportant fiber concrete, organic fibers such as coir, jute, cane splitting are also used.

  Applications : Sheet pipe, Board, Sewer pipe, Wall lining etc.

5. Carbon :

- Carbon fiber has very high tensile strength 2110 to 2815 N/mm^2 and Young's modulus. It has been reported that cement composite made of carbon fiber as reinforcement will have very high modulus of elasticity and flexural strength. Carbon fiber concrete is used to construct structures such as cladding, panels and shells.

3.10 SELECTION OF MATERIALS FOR REPAIRS

- Several repair materials are available commercially, their use depends upon nature and intensity of damage, type of environment, required life, importance of structure and available funds. Proper selection of repair materials is very essential for its effectiveness in repair.

  Following table facilitates in selecting the appropriate repair material for particular situation.

Table 3.6

Sr. No.	Type of Material	Advantages (Special Features of Material)
1.	Anti-corrosive coatings	• Anticorrosive steel primer. • Quick drying. • Two component mineral based corrosion inhibitor. • Can be welded through, forms conductive film. • Improves binding for further coats.
2.	Epoxy based adhesives (Solvent free)	• Bonding hardened concrete to fresh concrete. • Usable in wet conditions. • Develops high strength. • Has long open life. • Suitable for chloride contaminated host concrete.
3.	Water borne adhesive (Latex adhesives)	• Single component. • Water resistant bonding agent. • Polymer latex additive for repair mortar/concrete (PMC and PMM). • Fully compatible with PCC repair systems. • Universal carbonation resistance bond coat.
4.	Shotcrete admixtures (Sparayable repair mortars)	• Cementitious sprayable repair mortars. • Develops extra strength. • Applied fast. • High build/less rebound. • High abrasion resistance. • High chloride/carbon dioxide resistance.

5.	Prepacked cement mortars	<ul><li>They contain polymer to improve low permeability qualities and to enhance adhesion, heat development is low.</li><li>Suitable mortars for concrete repair.</li><li>Ready to use polymer based mortar for repair.</li><li>Provides increased bond strength.</li><li>Improves/provides resistance to carbonation.</li></ul>
6.	Epoxy resin mortar	<ul><li>Resin mortars suitable for small/shallow repairs.</li><li>High performance/Exceptional adhesion to concrete.</li><li>Better chemical resistance.</li><li>Rapid cure/Strength gain.</li><li>High strength-Abrasion impact-Tensile strength.</li></ul>
7.	Polyster resin mortars	<ul><li>Rapid setting- harder than concrete within 2 hrs.</li><li>No primer coat.</li><li>Good resistance towards bleach liquids.</li><li>High chemical resistance.</li></ul>
8.	Injection systems	<ul><li>Solvent free, two-part polyurethane for sealing cracks.</li><li>Low viscosity, high molecular weight polymer for grouting.</li><li>High bond strength.</li><li>Effective for structural repairs.</li></ul>
9.	Grouts	<ul><li>Innovative repair mortar.</li><li>Free flowing.</li><li>Non shrink.</li><li>High strength.</li></ul>
10.	Joint sealants	<ul><li>Elastomeric sealant for construction.</li><li>Excellent adhesion.</li><li>Accommodates high and low-temperature.</li><li>Accommodates high movement.</li></ul>

Important Points

- **Repair materials :** Selection of repair material is one of the most important tasks for ensuring durable and trustworthy repair. Analogy for selection of repair material is similar to repairing of a torn garment with sound fibre/fabric but of similar performance characteristics (i.e. preshrunk and similar fibre/fabric) as that of the original garment.

- **Essential parameters for maintenance and repair materials** are low shrinkage properties, Requisite setting/hardening properties, Workability, Good bond strength with existing sub-strate, Compatible coefficient of thermal expansion, Compatible mechanical properties and strength to that of the sub-strate, Allow relative movement, Minimal or no curing requirement, Alkaline character, Low air and water permeability, Aesthetics to match with surroundings, Cost, Durable, non-degradable or non-biodegradable due to various forms of energy, life, UV rays, heat etc. Non-hazardous/non-polluting.

- **Factors responsible for of Deterioration and Defects :** Mechanical agents, Electromagnetic Agents, Thermal agents, Chemical agents, Biological agents, **The Building Users. Earth Movement**

- **Waterproofing materials :** Traditional sealing and waterproofing systems include bituminous materials, plastic waterproofing foils and metal tapes for interior and exterior applications.

- **Repairing materials for masonry construction :** Non-shrink grout, polyester putty or 1 : 3 cement sand mortar, plastic/aluminum nipples, galvanized steel wire fabrics, clamping rods, wired nails, ferro-cement plates.

- **Repairing materials for RCC construction :** Materials for surface preparation, Chemical rust removers for corroded reinforcement. Passivators for reinforcement protection. Bonding agents, Structural repair Materials, Non-structural repair materials, Injection grouts, Joint sealants, Surface coatings for protection of RCC.

- **Special mortar :** Quick-Setting cement mortar, Air-entrained mortar, Polymer impregnated concrete, Partially impregnated and surface coated polymer concrete. Polymer cement concrete.

- **Fibre reinforced concrete :** Slurry infiltrated fibre concrete (SIFCON), Compact Reinforces Concrete (CRC).

- **Fibres :** Steel Fibres, Glass Fibres. Polypropylene Fibres, Asbestos, Carbon.

- **Aspect ratio :** The fibre is often described by a convenient parameter called "aspect ratio". The aspect ratio of the fibre is the ratio of its length to its diameter. Typical aspect ratio ranges from 30 to 150.

- **Type of paints :** Emulsion, Whitewash, Oil paint, Cement Based Paint, Enamel Paint, Distemper Paint, Bituminous Paint, Epoxy Paint, Anti-Condensation Paint, Luminous Paint, Latex Paint, Lead Paint, Metallic Paint, Rubber Paint, Aluminum Paint, Textured Paint, Silicone Paint, Zinc-Rich Paint, Anti-Corrosive Paint, Fungicidal Paint.

- **Emulsion paint has many benefits over other usual paint :** The surface of emulsion paint is hard and can be easily cleaned by washing with water. Exposure to water will not affect the colour or quality of emulsion paint.

- Emulsion paints are ideal for ceilings and walls because it is thick and easy to apply.

- Emulsion paints do not crack or fade in the sunlight.

- Emulsion paints hold well to almost any surface. It can also use without any pre-treatment.

- **Adhesive :** The adhesive used to bond the concrete and reinforcing material must be one that can ensure the required bonding strength. The adhesive used for the overlap splices for the reinforcing material must be one that can ensure the strength of the overlap splice section.

Questions for Practice

1. What is the action of shrink comb in expansive cement?
2. List the various types of polymer concrete.
3. Give the various monomers used in polymer concrete.
4. Define polymer concrete.
5. What are the uses of polymer concrete?
6. What is sulphur infiltrated concrete?
7. What are the applications of sulphur infiltrated concrete?
8. What is polymer impregnated concrete?
9. Define polymer partially impregnated concrete.
10. What is the difference between ordinary cement and expansive cement?
11. What is the use of corrosion inhibiting chemicals?
12. Write the use of anti-fungus admixtures
13. What are uses of curing compounds?
14. What are the uses of sealants?
15. What is the necessity for adding concrete chemicals?
16. What are the applications of Polymer cement concrete?
17. What is the criteria for selection for repair material.
18. What are the Essential parameters for maintenance and repair materials.

❑❑❑

Chapter **4**

MAINTENANCE AND REPAIR METHODS FOR MASONRY

Weightage of Marks = 12, Teaching Hours = 12

Syllabus

4.1 Causes of wall cracks due to bulging of wall, shrinkage, bonding, shear and tension, differential settlement of foundation, thermal movement and vegetation.

4.2 Probable crack location such as junction of main and cross wall, junction of RCC column and wall, junction of slab and wall, cracks in masonry joints.

4.3 Stages of repair : Material removal and surface preparation, fixing suitable formwork, bonding/passivating coat and repair applications.

4.4 Repair techniques : Grouting, patch spalling replacement or delaminating and epoxy bonded mortar.

4.5 Repairing methods for minor and medium cracks include epoxy injection, grooving and sealing, shotcrete, stitching, grouting and guniting.

4.6 Repairing methods for major cracks (width more than 5 mm) include fixing mesh across cracks, dowel bars, RCC band and installing ferro-cement plates at corners and propping.

4.7 Effects of dampness in wall, damping repair techniques such as replacement or inserting DPC in brick wall, bituminous painting, painting using water proof solution and cement with adhesive gum.

4.8 Causes and remedies of foundation settlement, improvement techniques by compaction, intruding sand piles, stone columns and grouting cement slurry.

Objectives

After learning this chapter, student will be able to

- Explain the wvarious causes of wall cracks and their probable locations.
- Select the relevant repair techniques for the damages in the given civil structures with justification.
- Explain the repairing methods for the different crack types for the given structure.
- Explain the damping effect and its repair techniques.
- Explain the various methods of improving the bearing capacity of foundation.

INTRODUCTION

- Buildings and other built structures are tend to wear and tear all the time, but usually this activity is so small as to be unnoticeable. This process can be caused by defects like bulging of wall, shrinkage, bonding, shear and tension, differential settlement of foundation, thermal movement and vegetation of the ground, foundation failure, decay of the building fabric, and so on.

- If a structure is unable to accommodate these above mentioned reasons cracking is likely to occur. The appearance of distortions and cracks can be visually unattractive and disconcerting, unsafe for occupants, and if left untreated they can affect the integrity, safety and stability of the structure. Effective treatment procedure requires first that, the causes of cracking are understood. Only then can a strategy for repair be implemented.

(4.1)

4.1 CAUSES OF CRACKS

- The **most common causes of cracking** are :
  - (a) Foundation failure due to the decay of soft clay brick, concrete erosion due to chemical contaminants, and so on.
  - (b) Ground movement (beneath foundations) caused by clay shrinkage, land slip, vibration, subsidence, settlement, heave, sway and so on.
  - (c) Decay of the building fabric, due to woodworm, rust and so on.
  - (d) Moisture movement that causes materials to expand or contract, perhaps due to the presence of vegetation or faulty or damaged drains.
  - (e) Thermal movement that causes materials to expand or contract as temperature increases or decreases.
  - (f) Suspended structures such as floors that deform under load.
  - (g) Tree root growth.
  - (h) Absence of foundations in older buildings.
- To avoid and minimize the cracks, first we should understand the basic causes, mechanism of cracking, and properties of the material used to construct the structures.
- **Following are the principle mechanisms causing crack in structures.**
  - (a) Moisture change.
  - (b) Thermal movement.
  - (c) Elastic deformation.
  - (d) Creep.
  - (e) Chemical reaction.
  - (f) Foundation movement and settlement of soil.
  - (g) Growth of vegetation.

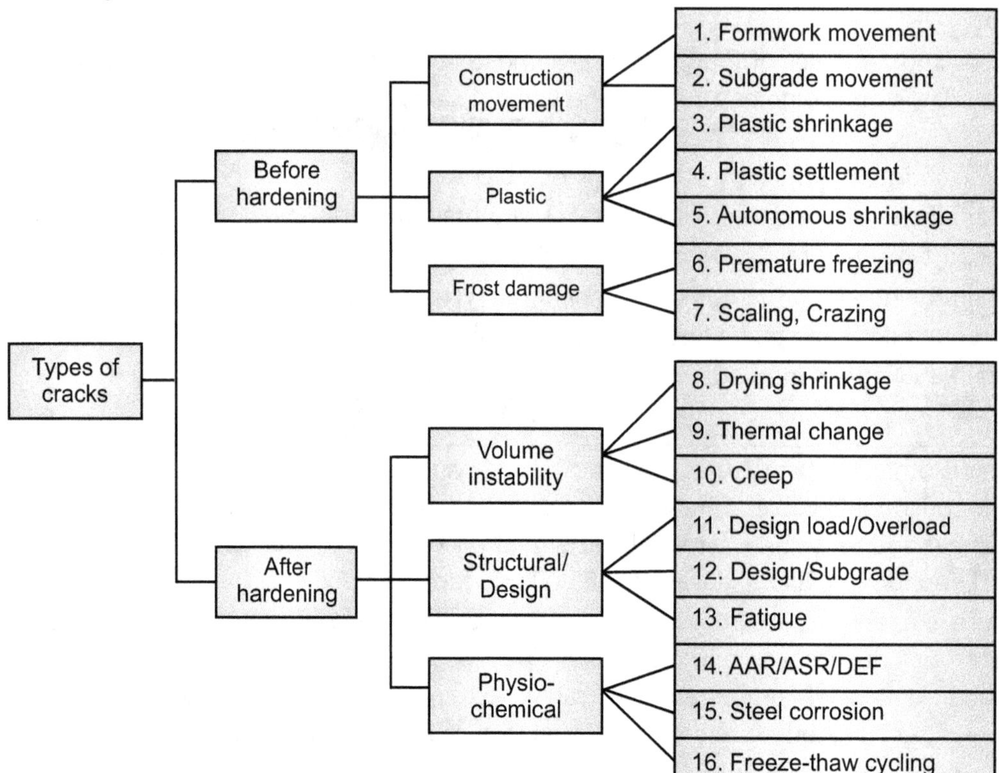

Fig. 4.1 : Common causes of cracking in concrete structures

4.1.1 Moisture Change

* Most of the building materials like concrete, bricks are porous in structure, These materials expands by absorbing moisture from atmosphere and shrinks on drying, it is reversible i.e. cyclic process happened due to inter pore pressure with moisture change. Extent of movement depends upon molecular structure and porosity of a material. Apart from reversible movement certain materials undergo some irreversible movement due to initial moisture changes after their manufacture or construction.

* The incidences of irreversible movement in materials are shrinkage of cement and lime based materials on initial drying i.e. initial shrinkage/plastic shrinkage and expansion of burnt clay bricks and other clay products on removal from kilns i.e. initial expansion. Initial shrinkage, which is partly irreversible, normally occurs in all building materials or components that are cement/lime based (e.g. concrete, mortar, masonry units, masonry and plaster etc.) and is one of the main cause of cracking in structure. Initial shrinkage of concrete and mortar occurs only once in the life time, i.e. at the time of manufacture/construction, when the moisture used in the process of manufacture/construction dries out. It far exceeds any reversible movement due to subsequent wetting and drying and is very significant from crack consideration.

* The extent of initial shrinkage in cement concrete and cement mortar depends on a number of factors namely :

    (a) **Cement content :** It increase with richness of mix.

    (b) **Water content :** Greater the water quantity used in the mix, greater is the shrinkage.

    (c) **Maximum size, grading and quality of aggregate :** With use of largest possible maximum size of aggregate in concrete and with good grading, requirement of water for desired workability is reduced, with consequent less shrinkage on drying due to reduction in porosity. e.g., for the same cement-aggregate ratio, shrinkage of sand mortar is 2 to 3 times that of concrete using 20 mm maximum size aggregate and 3 to 4 times that of concrete using 40 mm maximum size aggregate.

    (d) **Curing :** If the proper curing is carried out as soon as initial set has taken place and is continued for at least 7 to 10 days then the initial shrinkage is comparatively less. When the hardening of concrete takes place under moist environment there is initially some expansion which reduces a part of subsequent shrinkage.

    (e) **Presence of excessive fines in aggregates :** The presence of fines increases specific surface area of aggregate and consequently the water requirement for the desire workability, with increase in initial shrinkage.

    (f) **Chemical composition of cement :** Shrinkage is less for the cement having greater proportion of tri-calcium silicate and lower proportion of alkalis i.e. rapid hardening cement has greater shrinkage than ordinary port-land cement.

    (g) **Temperature of fresh concrete and relative humidity of surroundings :** With reduction in the temperature, the requirement of water for the same slump/workability is reduced with subsequent reduction in shrinkage. Concreting done in mild winter have much less cracking tendency than the concreting done in hot summer months. In cement concrete $1/3^{rd}$ of the shrinkage take place in the first 10 days, 1/2 within one month and remaining 1/2 within a year time. Therefore, shrinkage cracks in concrete continues to occur and widens up to a year period.

4.1.2 Plastic Shrinkage

* In the freshly laid cement concrete sometimes cracks occurs on the surface before it has set due to plastic shrinkage. Immediately after placing the concrete, solid particles tends to settle down by gravity action and water rises to the surface thus binding material of concrete and its aggregate gets separated. This process known as bleeding - produces a layer of water at the surface and this process continues till concrete has set. As long as the rate of evaporation is lower than the rate of bleeding, there is a continuous layer of water at the surface known as "water sheen".

- Shrinkage does not occur when the concrete surface loses water faster than the bleeding action bring it to the top, shrinkage of top layer takes place, and since the concrete in plastic state cannot resist any tension, cracks develops on the surface. These cracks are common in slabs. The extent of plastic shrinkage depends on :

(a) Temperature of concrete.

(b) Exposure to the heat from sun radiation.

(c) Relative humidity of ambient air and velocity of wind.

4.1.3 Thermal Movement

- All materials more or less expands on heating and contracts on cooling. When this movement is restraint, internal stresses are set up in the building component, and may cause cracks due to tensile or shear stress. Thermal movement is one of the most important causes of cracking in buildings. The extent of thermal movement depends upon :

(a) Temperature variation.

(b) Co-efficient of thermal expansion – Expansion of cement mortar and concrete is almost twice of the bricks and brick work. Movement in brickwork in vertical direction is 50% more than in horizontal direction.

(c) Dimensions of components.

4.1.4 Elastic Deformation

- Structural components of a building undergo elastic deformation due to dead and the super imposed live loads, in accordance with hooks law. The amount of deformation depends upon elastic modulus, magnitude of loading and the dimension of the component. This elastic deformation under certain circumstances causes cracking in the building as under :

  - When walls are unevenly loaded with wide variations in stress in different parts, excessive shear stress is developed which causes cracking in walls.

  - When a beam or slab of large span undergoes excessive deflection and there is not much vertical load above the supports (as in the case of roof slab), ends of beam /slab curl up causing cracks in supporting masonry.

  - When two materials, having widely different elastic properties, are built side by side, under the effect of load, shear stresses are set up at the interface of the two materials, resulting in cracks at the junction. Such a situation is commonly encountered in the construction of RCC framed structure and brick masonry panel (external) and partition (internal) walls.

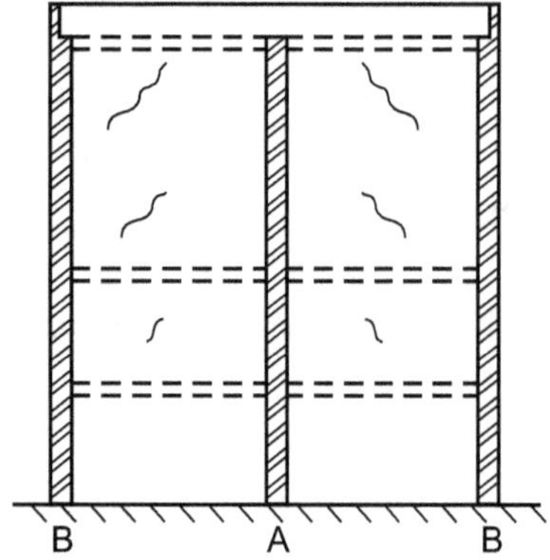
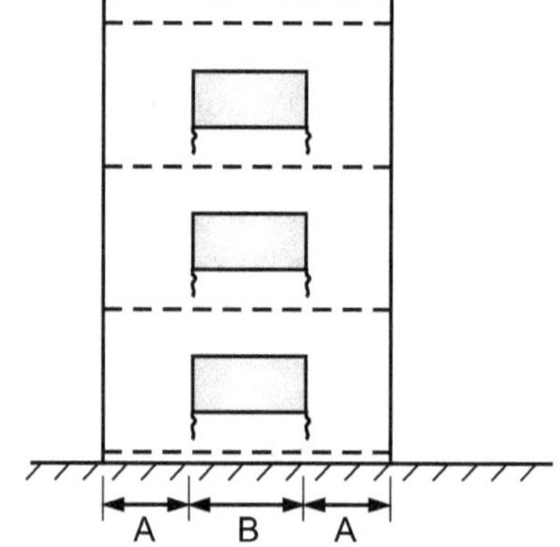

(a) Diagonal cracks in cross walls of multi-storeyed bearing structures   **(b) Vertical cracks in multi-storeyed buildings having window opening in load bearing wall**

Fig. 4.2

4.1.5 Creep

- Construction materials in general such as concrete, brickwork, mortar, wood etc. when subjected to constant loads undergo not only instantaneous elastic deformation, but also a gradual and slow dependence known as creep or plastic stress.

- In concrete, the creep extent depends on :
  - Water and cement materials.
  - Water cement ratio.
  - Temperature and humidity.
  - Use of admixture and pozzolana.
  - Age and strength of concrete at the time of loading.
  - Component size and shape.

- Creep increases with water and cement content, water-cement ratio and temperature. The creep decreases with increase in humidity of surroundings and the age/strength of the material at the time of loading. Use of admixture and pozzolana in concrete increases creep. Creep also increases with an increase in the surface in proportion to the amount of the component.

- In brickwork, creep depends on the stress/strength ratio so creep in brickwork with weak mortar is usually higher. For example : For the same quality of brick, creep in a 1 : 1 : 6 mortar is 2 to 3 times that of a brick work, which is 1 : 1 : 3 in a brick work. A simple creep in brick work is approximately 20 to 25% of concrete. In brickwork it stops after 4 months, while in concrete it can continue for a year or so, and most of the creep starts in 1 month after its speed slows down.

- The major effect of creep in concrete is a substantial increase in deformation of structural members, which can range from 2 to 3 times the initial elastic deformation. This deformation sometimes causes cracks in the brick masonry of the frame and load bearing structures. When the deformation due to elastic stress and creep is accompanied by a reduction of one RCC member due to shrinkage, the crack is much more severe and damaging.

4.1.6 Movement Due to Chemical Reaction

- Some chemical reactions in building materials result in appreciable increase in the material content, leading to internal stress setups that can result in external pressures and cracks. The material involved in the reaction also weakens in strength.

- Common examples of chemical reactions are :
  1. Sulfate attack.
  2. Carbonation in cement based materials.
  3. Corrosion of reinforcement in concrete and brickwork.
  4. Alkali-aggregate reaction.

4.1.7 Foundation Movement and Settlement of Soil

- Shear cracks in buildings occur when a large differential settlement of the foundation occurs due to one of the following reasons :
  (a) Uneven bearing pressure under different parts of the structure.
  (b) A pressure bearing greater than the safe bearing capacity of the soil.
  (c) Less safety factor in foundation design.
  (d) Local variations in the nature of the supporting soils which remained undetected and were not noticed in the design of the foundation at the time of construction.
  (e) Foundation resting in the active area on expensive land.

4.1.8 Growth of Vegetation

- The roots of a tree usually spread horizontally on all sides to the extent of tree height above the ground and when the trees are located around a wall, they can cause cracks in the walls due to the expansive action of the roots growing below the foundation.

- When the soil under the foundation of a building is shrinkable, cracks in the walls and floors of buildings can occur in the following ways :

  (a) Growing roots of trees cause de-hydration of soil which may shrink and cause foundation settlement.

  (b) In areas where a path was made for cutting down old trees so that the roots of the construction can be separated from the soil. On receiving moisture from some sources, such as rain, the soil swells and forces upward thrust on one part of the building, causing cracks in the building.

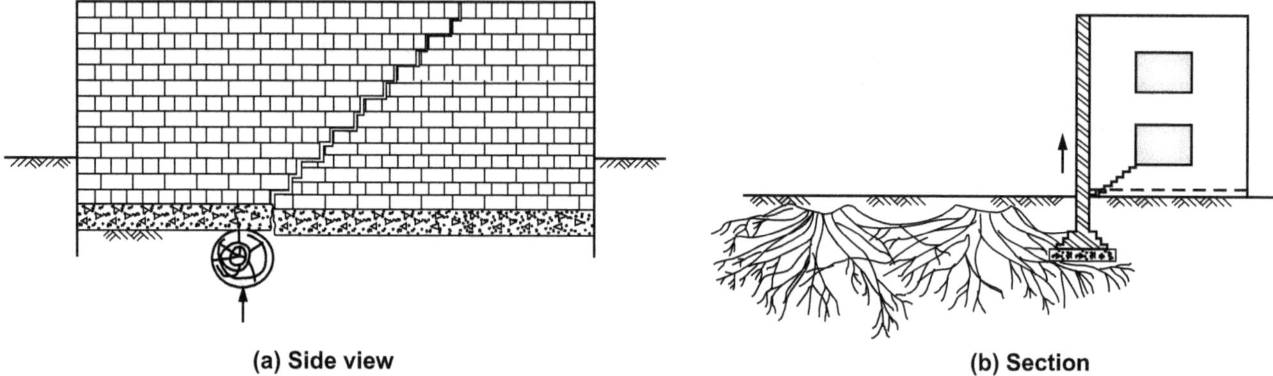

(a) Side view　　　　　　　　　　　　　　(b) Section

Fig. 4.3 : Cracks due to growth of vegetation

4.1.9 Choice of Materials

- For selecting materials for building construction following precautions shall be taken :

(i) Masonry units :

- Only well burnt bricks should be used for masonry.

- Burnt clay bricks and other burnt clay products should not be used in masonry for a period of at least 2 weeks in summer and 3 weeks in winter after unloading from kilns. They should be kept exposed to atmosphere during this period.

- Use of burnt clay bricks containing excessive quantity of soluble sulphates should be avoided if there use can not be avoided than rich cement mortar shall be used for masonry as well as plaster or super-sulphate cement shall be used. All possible steps shall be taken to prevent dampness in masonry.

- Use of porous stone with high drying shrinkage e.g. sand stone should be avoided for masonry and concrete work.

- While using manufactured masonry, units having high value of drying shrinkage for e.g. Concrete blocks and sand lime bricks, suitable precautions should be taken i.e. (i) They should be protected from wetting at site due to rains, and should be lightly wetted before use.

(ii) Mortar :

- The use of strong and rich mortar for laying should be avoided. Mortar used should have high water retentivity hence composite cement lime mortar are more preferred.

  (i) Curing of masonry should be done sparingly to avoid body of blocks getting wet.

  (ii) Before plastering, masonry shall be allowed to dry and undergo initial shrinkage.

  (iii) Excessive wetting of masonry at the time of plastering and curing should be avoided.

(iii) Fine Aggregate :

- Use of fine aggregate for mortar and concrete which is too fine or contains too much of clay or silt and is not well graded should be avoided. Percentage of clay and silt in fine aggregate (uncrushed) should not exceed 3%.

(iv) Cement :

- When use of bricks containing excessive quantity of soluble sulphates is unavoidable, content of cement in mortar should be increased or supersulphated cement should be used.

- If use of alkali-reactive aggregate is unavoidable, alkali content of cement should not exceed 0.6%. If low alkali cement is not economically available, use of pozzolanas should be made to check alkali-aggregate reaction.

- In massive structures, in order to limit heat of hydration, low-heat cement should be used.

- **Calcium Chloride :** Its use in concrete as accelerator should be avoided as far as possible. If unavoidable, its quantity should be limited to 2% of cement content.

- **Gypsum (Plaster of Paris) :** Gypsum plaster ($CaSO_4$) should not be used for external work, or internal work in locations, which are likely to get or remain wet. It should be remembered that gypsum and cement are incompatible, since in the presence of moisture, a harmful chemical reaction takes place.

(v) Steel reinforcement in brick masonry :

- Use of steel reinforcement in brick masonry should be avoided in exposed situations unless special precautions are taken to prevent rusting.

4.2 CRACKS IN BUILDING COMPONENTS

- Cracks in walls can be further grouped as :
  - (a) In masonry structure.
  - (b) In freestanding walls.

4.2.1 In Masonry Structure

- Commonly observed cracks in masonry structures are :

(i) Cracks at ceiling level in cross walls :

- In load bearing structures, where a roof slab undergoes alternate expansion and contraction due to temperature variation, horizontal cracks may occur (shear cracks) in cross walls, due to inadequate thermal insulation or protective cover on the roof slab.

(ii) Cracks at the base of a parapet wall :

- A frequency of thermal cracks occurring in buildings is the formation of a horizontal crack in support of a brick parapet wall or brick-less-iron railing on an RCC cantilever balcony.

- Factors, which contributes to this type of cracking are :
  - (a) Thermal coefficient of concrete is twice that of brick work and thus differential expansion and contraction cause of horizontal shear stress at the junction of the two materials.
  - (b) Drying shrinkage of concrete is 3 to 4 times that of brick masonry.
  - (c) Parapets are generally built over the concrete slab before the latter undergone its drying shrinkage fully, and
  - (d) Parapet or railing does not have much self-weight to resist horizontal shear force at its support caused by differential thermal movement and differential drying shrinkage.

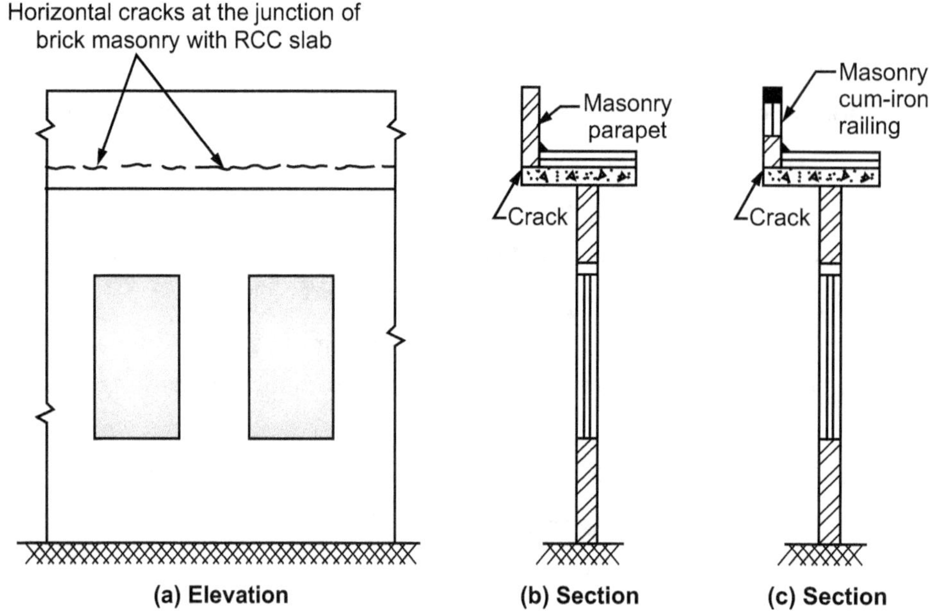

Fig. 4.4 : Horizontal cracks at the base of brick masonry parapet (or masonry cum iron railing) supported on a projecting RCC slab

(iii) Horizontal cracks in the uppermost storey below the slab level :

- These cracks are caused by deflection of the slab and lifting up the slab edge, combined with horizontal motion in the slab due to shrinkage. These cracks appear a few months after construction and are more prominent if the span is large. Due to the light vertical load on the wall, these cracks are confined to the topmost floor, due to which, the end of the slab is encountered without much restraint. In the lower storeys, vertical loads of the upper storey prevent the corners from lifting.

- Sometimes, due to deflection of the slab in both directions, lifting of the slab at the corners, horizontal cracks develop in the top floor of a building at the corners. **These cracks can be avoided** by providing adequate corner reinforcement in the slab.

- When large spans cannot be avoided, the slab or beam can be reduced by reducing the depth of the slab or beam to increase their stiffness. The adoption of special bearing arrangements (Figs. 4.5 and 4.6) and the provision of grooves in plaster at the junction of the wall and roof will help some in reducing cracks.

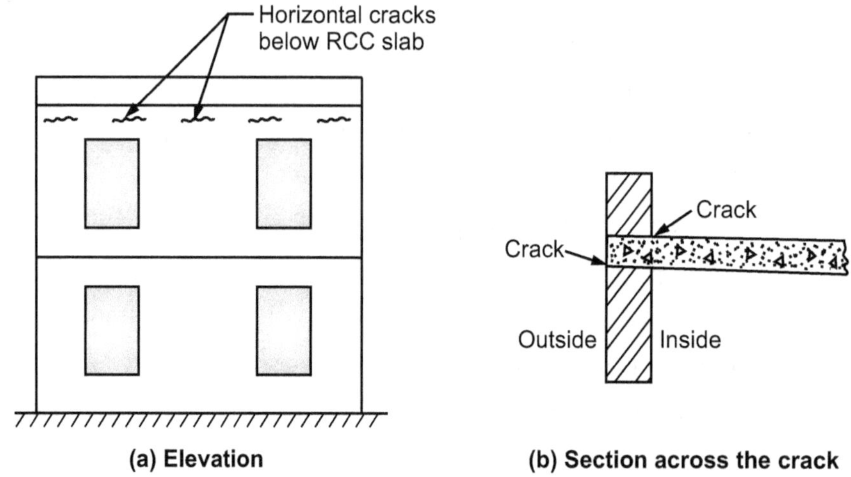

Fig. 4.5 : Horizontal cracks in top-most storey below slab due to deflection of slab

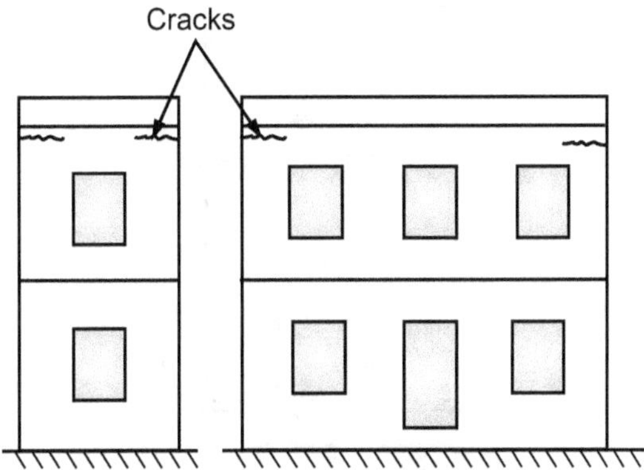

Fig. 4.6 : Horizontal cracks in top storey below slab level due to lifting of corners

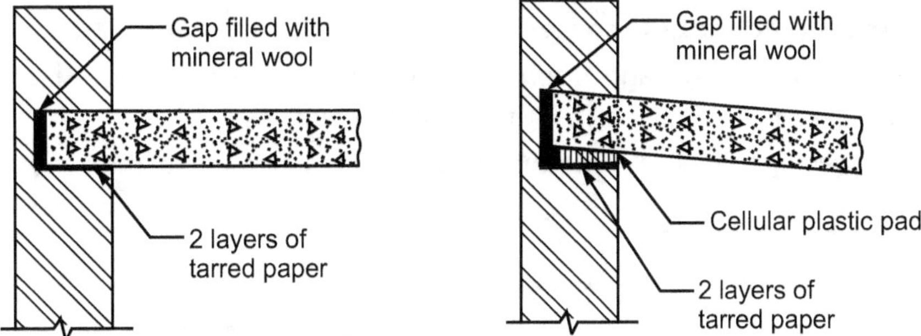

Fig. 4.7 : Details of bearing at the support for a roof slab

- **Diagonal cracks in the cross walls of a multi-storey load bearing structure :** These cracks are caused by differential stresses in the internal and external load bearing walls, thereby binding the cross walls. When the walls are unevenly filled with wide variations in stress at different parts, excessive shear stress develops which causes cracks in the walls.

- Fig. 4.8 shows, a multi storey load bearing structure with brick walls and RCC floors and roofs. When the central wall 'A', which carries more weight, the outer walls are 'B' and either the wall is the same thickness as 'B' or not correctly proportioned, it is higher than the walls 'B'. This causes shear stresses in the cross walls, which are bonded to the load bearing walls 'A' and 'B' and cause the diagonal crack as depicted in the Fig. 4.8..

- For the prevention of such cracks, it must be ensured that the stresses in the various walls of the load bearing structure are more or less the same in the design phase only.

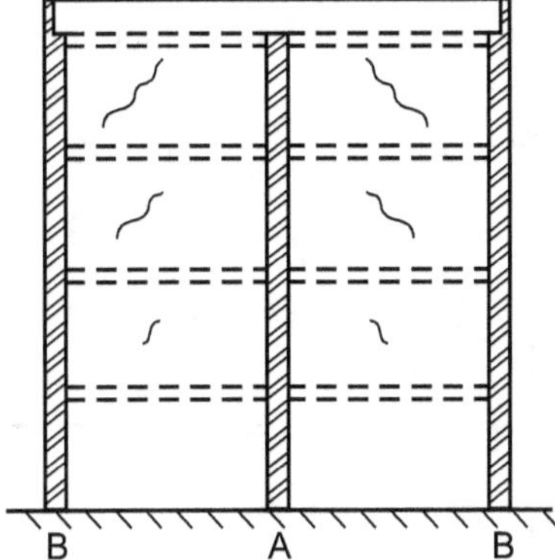

Fig. 4.8 : Diagonal cracks in cross walls of multi-storey load bearing structure

(iv) Vertical cracks below openings in line with window jambs :

- The height of another load-bearing multi-storey structure with large window openings in the outer walls is shown in Fig. 4.9. It can be seen that, parts of the wall are marked as 'A'. Under the pillars and windows there is much greater emphasis than in the parts marked 'B'. Thus, as a result of differential stress, vertical shear cracks appear in the wall as depicted in the Fig. 4.9.

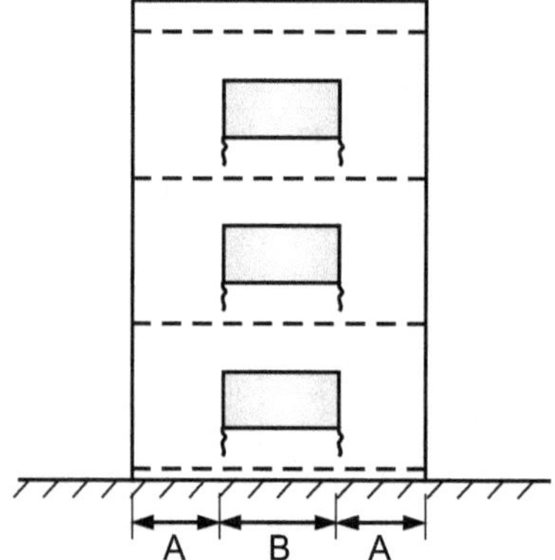

Fig. 4.9 : Vertical cracks in multi-storeyed building with window openings in load bearing wall

(v) Vertical cracks in the top floor of the corner :

- These cracks are caused by shrinkage upon initial drying of the RCC roof slab, as well as thermal contraction, which causes inward pull on the walls in both directions.

- These cracks can be prevented by providing the proper movement joint i.e. the joint between the slab and the supporting walls.

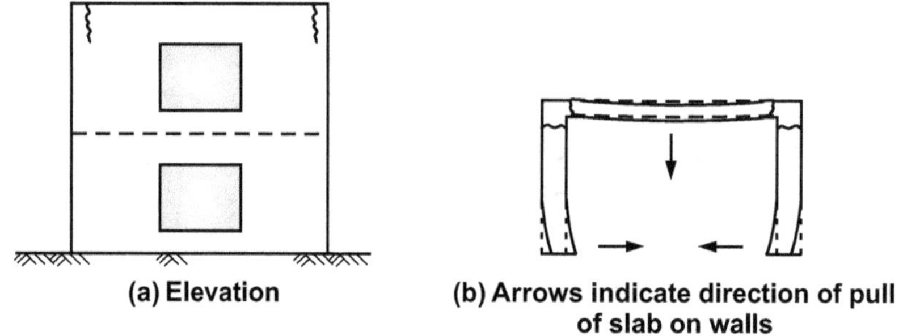

(a) Elevation   **(b) Arrows indicate direction of pull of slab on walls**

Fig. 4.10 : Vertical cracks at corners in the top storey of a building due to drying shrinkage and thermal contraction of slab

(vi) Vertical cracks around the stair case/opening of balconies :

- These cracks are caused by drying shrinkage/elastic shortening and thermal movement in the building. Generally, these cracks are not very obvious, and can be reduced by delaying the rendering/plastering to cause shrinkage/elastic deformation of the masonry/concrete.

(vii) Vertical cracks in the side walls at the corner of a tall building :

- These cracks are mainly due to thermal expansion, which increases with moisture expansion of the brickwork and is noticeable during hot weather. Such cracks occur more in a building built in cold weather. These cracks begin at the DPC level and go up to the ward, and are more or less straight and pass through masonry units.

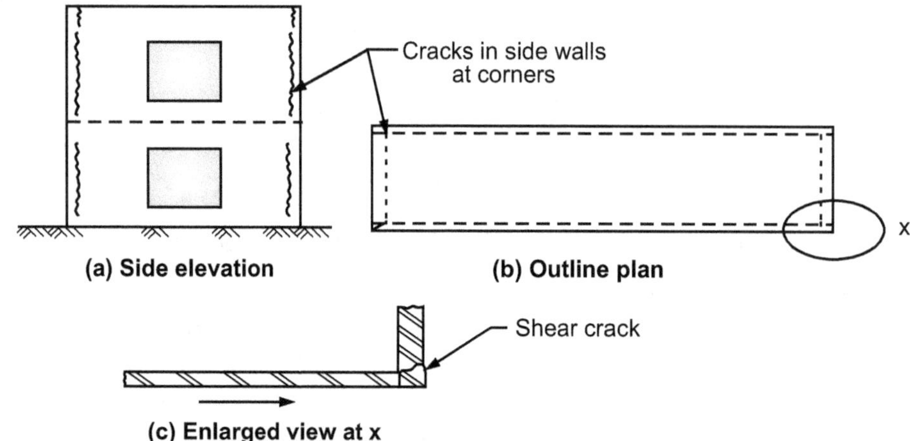

(a) Side elevation   **(b) Outline plan**

(c) Enlarged view at x

Fig. 4.11 : Vertical cracks at corners in the side walls of along building due to thermal movement

(viii) Horizontal cracks at the lintel/sill level in the top storey :

- These cracks occur due to drying shrinkage and thermal contraction caused by the slab being drawn to the wall. Such cracks typically occur when window and room spreads are large. These cracks can be prevented by providing slip joint on supportive measures to repair cracks in masonry walls.

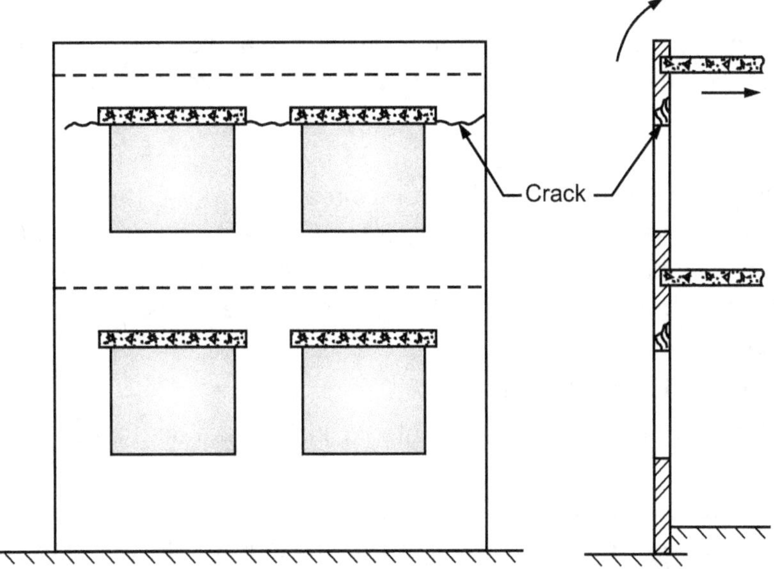

Fig. 4.12 : Horizontal cracks at window lintel level in top most storey

(ix) Diagonal cracks over RCC lintels spanning large opening :

- These cracks are caused due to drying shrinkage of in-situ RCC lintel and are observed during the 1st dry spell after construction. **These cracks could be prevented** by using low shrinkage and lows lump concrete or using pre-cast lintels.

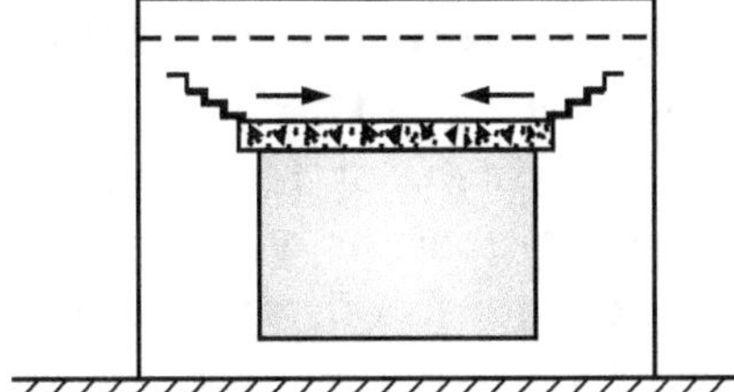

Fig. 4.13 : Diagonal cracks in brick masonry wall over lintel of long span openings

General Measures for Repairing Cracks in Masonry Walls :

- The main purpose of repairing cracks in the walls is as follows :

    (a) Restore normal appearance.

    (b) Reducing the possibility of cracks causing further damage to the building.

    (c) To ensure that the building is serviceable and safe.

- Before the cracks are taken for any repairs, it should be checked whether the cracks have stabilized and are not widening any further. Walls that do not exceed 25 mm from the plumb line, or that do not extend more than 10 mm in a typical storey height, will generally not require repair on structural ground. Due to thermal movement, the cracks usually reoccur when repaired with mortar. Therefore such cracks should be filled with some mastic compound.

    (i) Cracks up to 1.5 mm in width generally require no repair if the bricks are of the absorbing type, as is usually the case in India. In the case of non-absorbent bricks, rainwater is likely to penetrate through thin cracks, so it needs to be repaired in even thin cracks. The crack joints should be pulled out and filled with 1 : 1 : 6 (cement, lime, sand) mortar.

(ii) Cracks more than 1.5 mm wide usually require repair and the method depends on the type of mortar used in the brickwork.

    (a) With weak mortars, cracks should be enlarged and uprooted to a depth of about 25 mm and 1 : 2 : 9 (cement, lime, sand) mortar and replastered (about 10 cm wide strip around crack) with the same mortar.

    (b) With strong mortar, the bricks adjacent to the crack should be cut and new bricks replaced using 1 : 1 : 6 (cement, lime, sand) mortar. If the bricks are cracked then the same procedure should be followed.

(iii) Wide diagonal cracks, which are usually caused by the settlement of the foundation, if there is a possibility of further movement, should be repaired by removing and replacing all cracked bricks. RCC stitching blocks should be used in every 5th or 6th course, i.e. at a distance of about 0.5 m in the vertical direction. The stitch block should be equal in thickness to the wall thickness, half to 2 bricks in length and 1 or 2 bricks in thickness. It is not desirable to use a stronger mortar than 1 : 1 : 6 (cement, lime, sand) for these repairs.

4.2.2 Cracks in Free Standing Walls

- Following types of cracks are generally encountered in free standing walls, i.e. compound, garden or parapet walls :

    (i) Vertical cracks at regular intervals of 5 to 8 m and at change of direction may be due to drying shrinkage combined with thermal contraction. Cracks tend to close in hot weather. If wide enough, cracks may be repaired by enlarging them and filling the same with weak mortar (1 cement, 2 lime, 9 sand). If no expansion joints have been provided earlier, some of the cracks may be converted into expansion joints.

    (ii) Diagonal cracks, which are tapering and are wider at the top, are due to foundation settlement. If cracks are wide enough to endanger the stability of the wall, affected portion should be dismantled and rebuilt providing adequate foundation.

    (iii) Diagonal cracks, which are tapering and are wider at the bottom, may be caused by the upward thrust exerted by the roots of any trees and plants that may be growing in the vicinity of the wall.

    (iv) Arching up and cracking of the coping stone of a parapet or compound wall happens if the wall is built between two heavy structures, which act as rigid restraints, and no expansion joints have been provided in the coping stone. Remedy for this defect lies in relaying the affected portion of coping and providing expansion joint at suitable intervals.

    (v) Horizontal cracks in the bed joints of free standing walls, if the same occur two or three years after construction and the wall in question has been subjected to periodic wetting for long spells, may be due to sulphate action. This should be confirmed by chemical test of mortar and bricks. There is no effective remedy for these cracks and the damaged portion has to be rebuilt when it becomes unserviceable, taking other precautions for preventing recurrence of sulphate attack.

4.2.3 Cracks in Plastering and Pointing

(i) On masonry background :

- Cracks in plaster should be examined, if necessary, by removal of small portion of plaster, whether these are surface cracks or the see tend to the background masonry surface also. In the latter case, it is necessary to investigate the cause of cracks in the background material.

- If these are surface cracks, these could be due to :

    (a) Shrinkage because of use of rich mortar and inadequate curing - It occurs during 1st dry spell after construction.

    (b) Lack of bond with the background due to joints not being raked - Gives hollow sound on tapping the affected area.

    (c) Sulphate attack – Appears after 2 to 3 years after completion if the affected area remains damp for long spell.

Remedial measures for such cracks :

- Shrinkage cracks are generally thin and could be left as such up to the normal time for renewal of finishing coat of paint and distemper. The surface should then be thoroughly rubbed with emery paper No. 60 or 80 and there after two coats of paint may be applied, when the shrinkage cracks will get filled up.

- To repair cracks due to lack of bond, affected portion of plaster should be removed, joints in masonry raked to a depth of 10 mm and re-plastering done taking all precautions as required for a new plaster job.

- To repair cracks due to sulphate attack, source of dampness should be plugged. If mortar has become weak and un-serviceable, the affected portion should be re-plastered using sulphate resistant cement after removing all old mortar and raking joints in masonry.

(ii) On concrete back ground :

- Shrinkage cracks occur due to the following causes :
  - Mortar used being too rich or wet,
  - Curing has been inadequate,
  - Sand used is too fine, and
  - Rendering and plastering is done too long after casting of concrete.

- For prevention of such cracks, plastering should be done as soon as feasible after removal of shuttering, by hacking and roughening the surface and applying cement slurry on the concrete surface to improve bond.

(iii) Cracks around door frames :

- These cracks occurs due to :
  (a) Shrinkage of wood frames if the wood is not properly seasoned.
  (b) Door/window frames fitted flush with the wall/plaster surface.
  (c) Due to slackness between the hold-fasts and frames - Heavy vibrations causes crack in masonry and plaster due to repeated opening and closing of doors.

- Cracking due to in-proper seasoning can be concealed with the help of architrave, however for repair the cracks due to slackness in hold-fast masonry is to be dismantled to remove the frame and re-fixing it after securely fastening the hold-fasts to the frame.

- When the doors/windows are provided in half brick masonry wall, hold-fast should be at least 25 cms in length and should embedded in 1 : 2 : 4 concrete, in 2-brick courses in height and $1\frac{1}{2}$ brick in length.

- To prevent such cracks, as far as possible, door and window frames should not on either side be fitted flush with wall/plaster surface. Where unavoidable, either conceal the junction with architrave or provide the frame of shape and design as shown in Fig. 4.14.

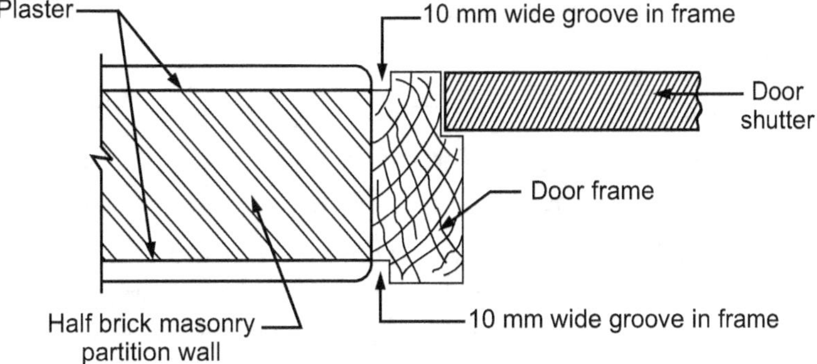

Fig. 4.14 : Arrangement showing fixing of door frame flush with wall surface

4.3 STAGES OF REPAIRS

- Before carrying out any repairs to cracks, it should be examined whether the cracks have stabilized and are not widening any further. Walls which are not more than 25 mm out of plumb, or which do not bulge more than 10 mm in a normal storey height, would not generally require repairs on structural ground. Cracks due to thermal movement generally reoccur when repaired with mortar; therefore such cracks should be filled with some mastic compound.

- Masonry consists of stone, brick or concrete blocks bonded with mortar. Mortar is a mixture of lime, sand, water and, more recently, cement. Masonry, when properly constructed and well maintained, will last for centuries. Mortar joints will normally require re-pointing every 50 years. Brick is the most common masonry wall, usually with limestone or granite for foundations.

- Stone is also common in areas where there is a tradition of stone masonry and availability of the material. The performance of a masonry wall depends on proper design, materials, workmanship and maintenance. Key requirements are that a wall be strong, keep moisture out and allow the mortar to flex. Most masonry work involves re-pointing and occasional replacement of deteriorated brick or stone.

4.3.1 Deterioration of Masonry

- Brick and stone can deteriorate for many reasons, the key ones being :
  - (a) Excessive moisture in the masonry freezing and thawing in winter.
  - (b) Water in the walls rusting out masonry ties.
  - (c) Improper cleaning, such as sandblasting.
  - (d) Differential expansion, leading to cracking.
  - (e) Airborne pollutants.

Some indications of masonry problems are :

Bulging	-	indicates the wall has moved.
Cracking	-	indicates movement within the wall.
Straining	-	indicates excessive dampness.
Crumbling	-	indicates moisture penetration due to poor brick or to sandblasting.
Paint Blistering	-	indicates moisture trapped behind paint.
Mortar Cracking	-	indicates cement mortar is too hard and is popping out in freeze-thaw cycles.

Principle :

- (a) Conserve original brick and stone with periodic check-ups and maintenance.
- (b) When re-pointing, use an appropriate lime-based mortar.
- (c) Avoid cleaning. Conserve the patina of age that gives the building character.
- (d) If cleaning is necessary, carry out with great care. Do test patch first to ensure there is no damage to the masonry surface.
- (e) Never sandblast.

- **Main purpose of carrying out repairs to cracks in walls** is to :
  - (a) Restore normal appearance.
  - (b) Minimize the possibility of cracks causing further damage to the building.
  - (c) To ensure that the building is serviceable and safe. Before carrying out any repairs to cracks, it should be examined whether the cracks have stabilized and are not widening any further.
  - (d) Walls which are not more than 25 mm out of plumb, or which do not bulge more than 10 mm in a normal storey height, would not generally require repairs on structural ground. Cracks due to thermal movement generally reoccur when repaired with mortar; therefore such cracks should be filled with some mastic compound.

4.3.2 Precautions During Repairs

Re-pointing :

- The mortar bonds masonry together, at the same time, compressive strength, workability and flexural (bending) capacity are important. The mortar walls must be weakened by masonry to accommodate movement, otherwise cracks will occur. Earliest the mortars were lime based, water resistant and flexible, but often vulnerable and susceptible to frost action. Later cement mortars, with little or no lime, are strong and fast setting, but with poor and uncertain bonding. Masonry cement is a pre-mixed mixture of lime, portland cement, and other ingredients that may vary depending on the circumstances.
    - Duplicate the original mortar mix.
    - Match the original mortar joint.
- In applying the mortar, ensure that the adjacent bricks are wet and the mortar dries slowly under the shade of a tarpaulin if it is sunny. Allow it to cure properly.

Mortar Mixes :

- The correct mortar mix for a masonry wall is very important, so that, it moves with the wall. If the original mortar has performed well, the intent is to duplicate its mix.

Cleaning :

- Cleaning masonry is one of the most difficult jobs, particularly when trying to remove paint. In general, there is no way to remove paint that will not damage the surface of the masonry. The only solution is to allow the paint to deteriorate or surpress over many years. Cleaning dirt off masonry is a simple and safe process, although dirt does not harm masonry and in fact provide attractive qualities of character and age.
- Cleaning methods include the following :

(a) Water :

- Cleaning masonry with water is the simplest, safest and least expensive method. It softens the dirt and rinses deposits from the surface. When water-cleaning, ensure the wall is watertight and mortar and caulking joints are sound, the least amount of water is used, and there are two to five weeks of dry weather before frost.
- The different techniques are as follows :
    1. Hand-scrubbing - using a mild detergent and hosing down when complete. This is simple and effective.
    2. Spraying - using regular water pressure to create a fine mist applied periodically over several hours and hosing down when complete.
    3. Pressure washing - Using mechanized pressure. Great care should be taken on soft masonry and mortar, which can be destroyed if the pressure is too high and spray duration too long.

(b) Chemical :

- Chemicals are usually used to remove paint. It can, and usually does, destroy the surface of masonry. If contemplated, a test patch should be done to determine the extent of the damage. The general approach to chemical cleaning involves wetting down the masonry, applying the chemical and rinsing off.
- The different cleaners are as follows :
    1. Acid - usually hydrofluoric (HFI), is mixed in a maximum concentration of 5%, preferably 1%-3%. Acid should not be used to clean limestone, marble or sandstone.
    2. Alkali - can be used on acid-sensitive masonry such as limestone, marble and glazed brick. It has a potassium hydroxide, ammonia or caustic soda base. Alkali should not be used on stone with a high iron content.
    3. Paint removers - are often the only means of removing paint. Reaction with the masonry can vary, therefore a test patch should be conducted first.

(c) Sandblasting :

- Abrasive cleaning, usually sandblasting, is not acceptable for old and historic masonry. It removes the hard exterior surface of brick, in particular, which then take on moisture and rapidly deteriorates. Many older brick buildings which were sandblasted have subsequently been re-plastered as the brick became porous and crumbled. On stone, it can destroy details and texture.

4.3.3 Repair Stages in Masonry

1. Repair Methods for Cracks in Masonry Building Structural Members :

- Measures to be followed for already appeared cracks are :

   1. Application of grouting or uniting to cracks appearing in the main structural members that cannot be compromised at any cost. The material used for this is either cement or epoxy mixture. The epoxy has the ability to fill small and thin cracks, say as fine as 0.1 mm. These epoxy achieve high strength and adhesion.

   2. The flexible sealant can be used for cracks visible on the non-structural members. This helps in controlling the differential movement (expansion or contraction) of the member under temperature changes.

   3. Epoxy putty, polymer filler or lime cement mortar can be used for filling the cracks seen in plain cement concrete.

2. Measures for Foundation Settlement :

- The uneven settlement of foundation due to the variation of bearing capacity at different points of the building results in cracks in the building.

- Some preventive measure are :

   1. The foundation is planned to lay or hard soil.

   2. The foundation and wall have to be raised slowly, keeping the structure for allowable settlement.

   3. The settlement value should not go beyond allowable, under any combination of loads.

   4. The designed foundations should facilitate uniformly, distributed pressure on the soil.

3. Plinth Protection :

- The uneven settlement of plinth avoids removal of expansive soil such as black soil (black cotton soil), nearby plinth. This barrier is kept with the help of sand piles.

- Providing drains and flagging concrete help to avoid rainwater away from the plinth.

- The penetration of roots into the plinth has to avoided. This can be avoided by stopping the construction of trees with roots growing nearby.

- Cracks 1 and 5 can be repaired by the crack injection process. In addition to the repairing the crack, framing supported on the adjacent wall around the corner should be examined to check for lack of support or damage, and appropriate repairs should be made. It is recommended that shear connection of the diaphragms to the wall also be made.

- Crack 2 can be repaired by the injection process. Adjacent to a steel lintel at the top of the opening, injection from both sides of the wall is required because the lintel will prevent injection from one side to the thickness of entire wall.

- Cracks 3, 4, and 6 will generally be closed by gravity and cannot receive injected grout. Grout can be used to fill the collar joints adjacent to the crack and bond adjacent loosened bricks.

- Crack 7 cannot be repaired by injection. The arches were generally constructed with special effort to completely fill collar joints and head joints. Voids are small and not interconnected so that the injection process is not effective. The portion of a badly displaced arch with or under loose brick will be reconstucted.

- Crack 8 do not have the typical pattern of bed and head joints. The loosened portion is separated from the wall along a diagonal line that has sheared through bricks. Restoration of support for the lintel requires that, the corners of the pier be reconstructed after removal of the damaged bricks.

- Crack 9, up to about 3/4 inch in width can be repaired by injection process under the following conditions :

  - Related cracks similar to crack 8 shall be repaired as noted for crack 8.

  - Where the cracks form a part of the walls in lateral load protecting system, the bed joints on which sliding has occurred shall be raked to a depth of 1 ½ inch and repointed with mortar at both the interior and the exterior faces. The building shall be analyzed and the overstress that caused shear cracks in the pier shall be addressed.

  - Crack 10 that is related to foundation settlement or causes other than earthquakes may be repaired by the insusceptible jection process after the causes have been addressed.

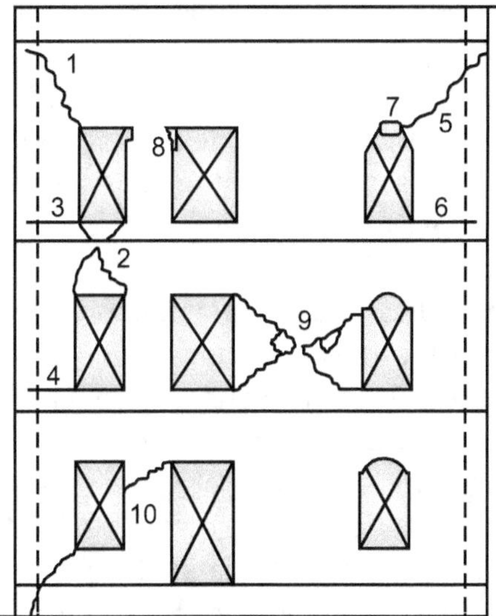

Fig. 4.15 : Remedial measures for dampness and efflorescence in wall

4. Removing Efflorescence :

- Efflorescence consists of white deposits on the brick surface left when moisture carrying dissolved salts evaporates.

- Prior to cleaning efflorescence, potential water penetration issues in the area should be investigated and resolved; otherwise, the efflorescence will return. Any leaks should be repaired and the brickwork allowed to dry. Generally, efflorescence is water soluble and easily removed by natural weathering, by dry-brushing or by scrubbing with a stiff fiber brush and water. Proprietary cleaners formulated specifically for use on brickwork are effective in removing heavy accumulations or stubborn efflorescence.

- Use solutions specifically manufactured, to remove efflorescence from brickwork. Improper cleaning procedures such as insufficient prewetting, insufficient rinsing and strong chemical concentrations may cause additional staining, engraved mortar joints and increase moisture penetration in brickwork. Stains caused by improper cleaning are not water soluble but can be removed by proprietary cleaners. To avoid improper applications of proprietary cleaners, it is necessary that the manufacturer's instructions be carefully followed.

- All cleaning processes must be tried at different concentrations in an ordinary area to judge their effectiveness and potential harm to the brickwork prior to implementing at full scale. After cleaning, the mortar joints should be inspected. Repointing or grouting of the joints, may be necessary.

5. Installation of a Damp-proof Course :

- The movement of moisture to a wall through the brickwork by capillary action is referred to as rising damp. This condition appears as a rising water line or "tide mark" on the wall caused by the soluble salts in the groundwater. Consequently to reduce the potential for rising damp, a damp proof course should be installed.

- Model building codes require the use of a damp proofing or waterproofing material on the surface of masonry walls located below grade and to install base flashing to be installed within 10 in. (254 mm) of final grade. If these are omitted or improperly installed, then rising damp may occur. The insertion of a damp proof course at a level above the ground, but below the first floor, may prevent the rising moisture.

- There are different methods to install a retrofit damp proof course. One method involves injecting a synthetic chemical that forms a continuous damp proof barrier into an existing brick course. Holes are drilled into the course of brick, and the synthetic material is injected. Another method is the creation of an additional flashing level just above grade. One or more brick courses are removed, flashing is installed and the brick are replaced. Recommendations for brick removal and replacement are discussed in the preceding section. In situations with severe rising damp, excavation of the soil adjacent to the wall and installation of damp proofing or waterproofing materials to the wall surface may be required.

6. Installation of Flashing :

- Building codes require flashing to be designed and detailed to resist water penetration to the interior of the building. Flashing that has been omitted, damaged or improperly installed may permit moisture to penetrate to the building interior. If this is the case, then flashing can be repaired or replaced to correct uncontrolled water penetration related to the flashing system. The process is aggressive, requiring the removal of brick, brazing the brick above, installing new flashing and replacing the removed brick units.

- Other methods may be used to address water penetration, but these are not necessarily long-term solutions and will not comply with the building code when flashing is missing. To install continuous flashing in existing walls, alternate sections of masonry in 2 to 5 ft (610 mm to 1.52 m) lengths should be removed. The replaced masonry should be properly cured (five to seven days) before the intermediate masonry sections or supports are removed. Alternately, temporary braces can be installed if longer sections of brickwork are removed. After these braces are installed, the flashing can then be placed in these sections. The lengths of flashing should be lapped a minimum of 6 in. (152 mm) and the laps adhered and edges sealed with a sealant or adhesive compatible with the flashing material to function properly.

4.3.4 Walls and Columns

- Concerning walls, to counteract the effects of vertical loads, the most efficient measures are the consolidation of the material itself, through injection or repointing, as has been referred to for the masonry material. In the case of composite walls, with two exterior skins and an interior core (usually, rubble of low quality), steel connectors, anchored in the exterior skins constitute an efficient measure to assure their integrity, impeding their separation from the interior core. When the walls are cracked due to in-plane loading, an adequate solution will be the installation of anchorages (anchor bolts) in the thickness of the wall, crossing the cracks. In particular situations, it can be adequate to simply clamp the cracks with appropriate clamps, anchored on the surface of the wall.
    (a) With anchor bolts.
    (b) With clamps in cracks.
- To counter the effects of lateral loads on walls they can be strengthened with metallic reinforcement or strips of composites : carbon fibres (CFRP), glass fibres (GRC) etc., applied on their faces. In some situations, strutting can also be adequate.

- Concerning columns, injection, the application of ties and lateral confinement by wrapping with metallic sheets or composites (CFRP, GRC etc.) are the measures usually used. The best solution will depend on the specific conditions.

4.3.5 Arches, Vaults and Domes

- For this type of element, in addition to the injection and the re-pointing (as referred to for walls), other measures can be adopted, such as the introduction of tie-rods, generally, in steel, to compensate for the thrust induced on the supports.

- The tie-rods should be placed, preferably, at the level of the bearing (in arches and vaults), or along parallel circles (in domes). They should be installed with a slight degree of pre-stressing, in order to guarantee that they will always be under tension.

- When it is possible to install, another solution will be the jacketing of the extrados of the vaults with strips of glued composite materials. However, in this case, attention should be paid to the barrier effect that is created along the vault. An alternative solution, very useful in particular situations, will be the insertion of strengthening components (in steel or timber), glued to the extrados of the masonry, which provides stiffness to the vault and in which the barrier effect is much less sensitive.

- When blocks have become out of position, an adequate solution will be dismantling, followed by rebuilding in the correct position. In very severe situations, when the shape of the element has been heavily changed, it can be more appropriate to demolish the element, followed by its reconstruction using materials similar to the original ones.

- In the case of filled vaults, one possible solution will be the reduction of weight or, if appropriate, the adjustment of its distribution.

- In the case of floors made with brick vaults supported on steel beams, an adequate measure to restrict their lateral separation will consist in the placing of tie-rods, in steel, welded onto the bottom flange of the beams.

4.3.6 Towers and Chimneys

- The most common solution for the strengthening of this type of element consists of tying them with horizontal ties, usually, steel strips or cables, or by wrapping them with glued composites (CFRP, GRC, etc.). In the case of prismatic towers an appropriate measure will be the provision of diaphragms (in concrete or steel), at intermediate levels, for the confinement of the walls.

4.3.7 Framed Elements

- In the case of timber elements, the most common solution for their strengthening is the substitution of the damaged elements by new ones. Sometimes, the existing members can be strengthened, for example, through the gluing of strips or wraps of composites : CFRP etc. In the case of steel elements the best solution will also be the substitution of the damaged members by identical new ones, eventually stronger.

4.3.8 Wood Siding

- Wood siding was one of the most common sidings for historic buildings. Wood is vulnerable to decay through moisture and damp and requires a finish of paint or stain. Every effort should be made therefore, to preserve original wood siding, not only because it is authentic to the building but also because it's quality cannot be found. It is the deterioration of the finish that has frequently caused wood siding to be covered by insulated brick in the early 1900's and, more recently, by vinyl or aluminium siding. The main challenges of wood siding are, therefore, maintenance or choosing the right replacement.

4.3.9 Cracks Prevention

- Non-structural cracks in buildings usually occurs due to more than one cause as already mentioned in previous chapter, therefore measures for prevention of cracks in many cases are common to more than one cause. Measures for prevention of cracks could be broadly grouped under the following sub-heads :
  - Choice of materials,
  - Specifications for mortar and concrete,
  - Design of buildings (architectural, structural and foundation),
  - Construction techniques and practices, and
  - Environment.

4.4 REPAIR TECHNIQUES

4.4.1 Grouting

- Grouting is a process of filling the cracks, voids under pressure in masonry structural members for repairing of cracks, strengthening of damaged masonry structural members.
- Grout is a flowable plastic material and should have negligible shrinkage to fill the gap or voids completely and should remain stable without cracking, de-lamination or crumbling.

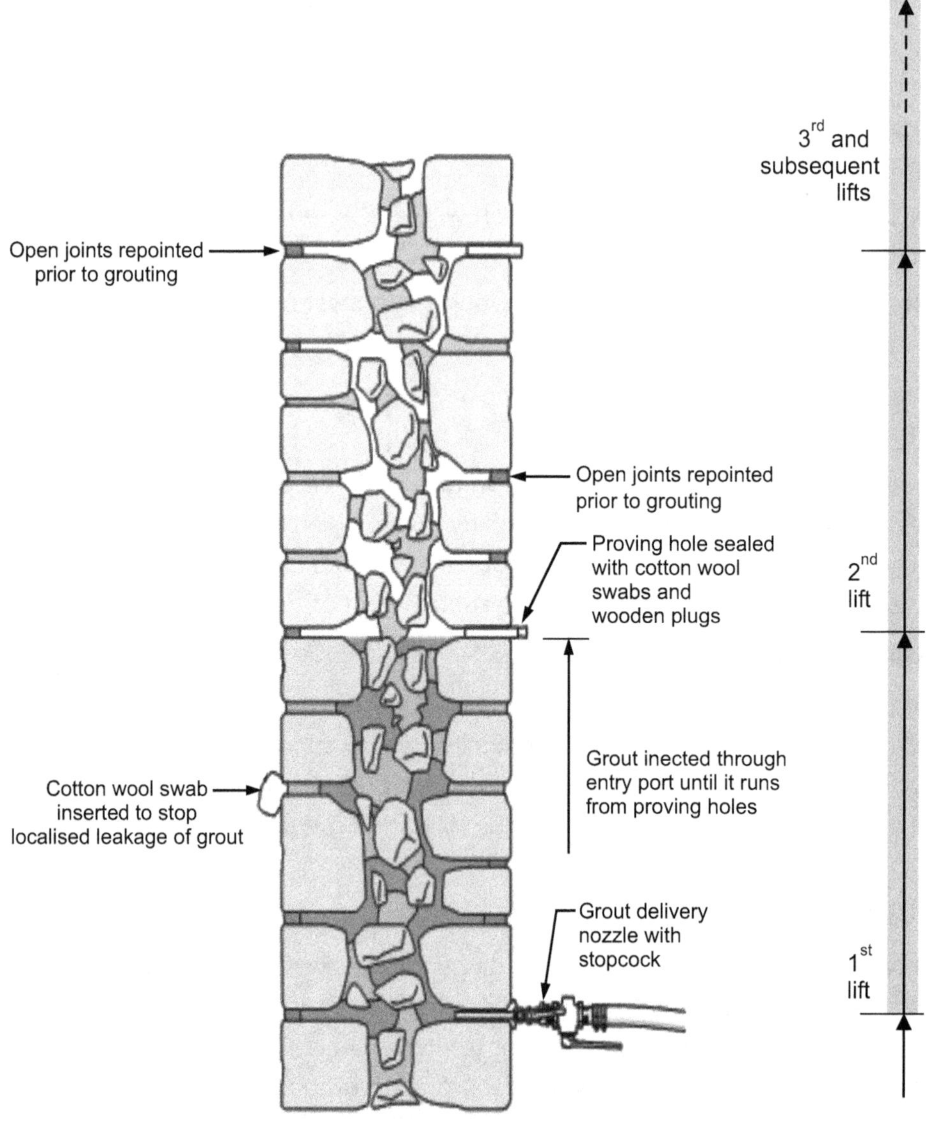

Grounting voids in a composite wall

Fig. 4.16 : Types of injection grouting and materials

- There are different types of grouts used for repair and strengthening of masonry structural members. The selection of type of grout for particular type of masonry repair work should be based on the compatibility of the grout with the original material.

- Following are the various types of injection grouting and materials :

1. **Polymer Injection Grouts :**

- Polymer injection grouting is based on use of polymers such as polyester, epoxy, vinyl ester, polyurethane and acrylic resins. The most popular polymer used for epoxy grout is epoxy. Polyurethane and acrylic resin based polymers are used for treatment of water retaining structure, underground structures as well as to prevent seepage of water.

- The polymer injection grouts are available in three component materials and two component materials. The three component materials grout includes :

   (i)　Liquid resin content,

   (ii)　Curing agent or hardener,

   (iii)　Aggregate or dry filler,

The two component materials grouts include :

   (i)　Curing agent or hardener

   (ii)　Aggregate or dry fillers.

- The polymer injection grouts can be made suitable for repair works by adding modifiers to basic resins and curing agents to achieve the desired properties. These vary from manufacturer to manufacturer, and therefore should always be used based on manufacturer's recommendations.

Following are Properties of Different Types of Polymer Based Injection Grouting :

Epoxy based injection grouts possess low pot life, non-resistant to ultraviolet exposure and high temperatures, non-shrink, flowable, effective in sealing cracks, excellent bonding with almost all building materials, good chemical resistance.

Acrylic polymer based injection grouts possess improved flexural and tensile properties, resistance to cracking, segregation, improved imperviousness, chemical resistance, rapid setting. Shrinkage can reduce/increase steel corrosion, dynamic load/vibrations resistant.

Lignosulfonate-based injection grout admixture lowers viscosity of cement slurry, compensates drying and plastic shrinkage.

2. **Fiber-Reinforced Injection Grouts :**

- Fiber reinforced concrete is used for repair of concrete and masonry structural members. Fibers such as polypropylene, steel or glass fibers are used with portland cement or shrinkage compensating mortar to repair and strengthening of structural members to provide improved flexural strength, impact resistance and ductility. Fiber reinforced injection grouts require skilled handling to avoid segregation of fibers.

3. **Cement–Sand Grouts :**

- Cement-sand grout is the most popular type of grout used for repair of concrete or masonry structure and is easily available. This grout is used for the places where strength enhancement of structure is not required. This is also most popular because it is readily available in the market and is cheapest form of repair of concrete and masonry structural members.

- This method requires high water and cement contents for injection purpose. The use of cement-sand grouts results in shrinkage and cracking of grout at hardening and to minimize this, suitable shrinkage compensating agents are required. Use of cement-sand grouts is very common in masonry buildings, but not very common in concrete.

4. Gas-forming Grouts :

- The gas-forming injection grout is used based on the principle that the gas bubbles expand the grout to compensate shrinkage of grout after application. These gas bubbles are generated on reaction of some ingredients (usually Aluminium and Carbon powder contained in grout) with the cement liquor.

- The gas bubble forming injection grouts are sensitive to temperature and are not suitable for high temperature application, require proper confinement to develop strength and volume stability, as the reaction forming gas bubbles may be too fast and may complete before placing of the grout.

5. Sulfo-aluminate Grouts :

- Sulfo-alum injection grout is also based on the principle shrinkage compensation. In these grouts either shrinkage-compensating cement or anhydrous sulfo-aluminate expansive additive is used with Portland cement. The additive results in expansion at hydration.

- This produces expansion after the grout has set and is more reliable than gas-forming grouts. But the expansion of such grouts requires post-hardening curing and it will not be effective if moist curing is not available.

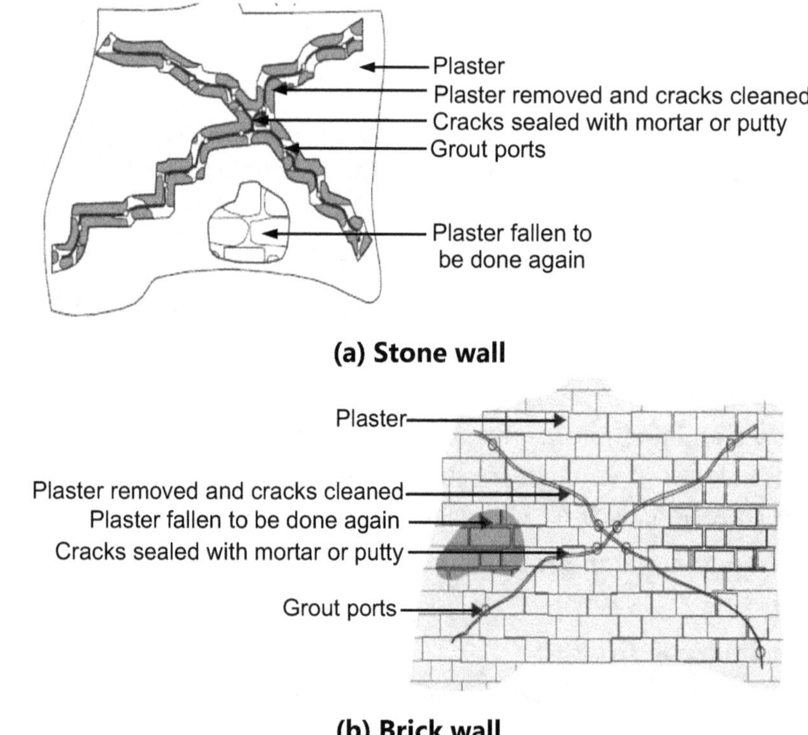

(a) Stone wall

(b) Brick wall

Fig 4.17: Grouting in masonry

4.4.2 Patch Spalling Replacement

- Where an area of plaster has become detached from masonry it can sound hollow when tapped. Plaster, especially excessively thick layers, is inclined to debond from the wall when the outside skin shrinks at a different rate than the plaster in contact with the wall. This is generally the result of inadequate substrate preparation.

Cutting of Old Plaster:

- The mortar of the patch, where the existing plaster has cracked, crumbled or sounds hollow when gently tapped on the surface, is first removed. The patch is be cut out to a square or rectangular shape at position where repairing is needed. The edges of cut plaster is made under cut to provide a neat joint.

Preparation of Surface:

- The masonry joints which become exposed after removal of old plaster is raked out to a minimum depth of 10 mm in the case of brick work and 20 mm in the case of stone work. The raking is carried out uniformly with a raking tool, and loose mortar is dusted off. The surface is then thoroughly washed with water, and kept wet till plastering is commenced.

- In case of concrete surfaces, the old plaster is thoroughly scrubbed with wire brushes after the plaster had been cut out, and pock marked the surface is roughened by wire brushing, and all the resulting dust and loose particles cleaned off. The surface is washed and cleaned and kept wet till plastering is commenced.

Application of Plaster :

- Mortar of specific mix such as 1:4 or 1:6 with the good quality plaster sand is used. After the plaster has been applied to the surface, finishing of plaster is done to match with the old surrounding plaster.

Curing and finishing of Plaster:

- Curing of plaster is necessary to prevent cracking. It should be done for at least 3 days at regular interval. After the plaster is thoroughly cured and dried the surface is then painted with the color of the surrounding area.

4.4.3 Bonding Plaster

- Bonding is an undercoat plaster. This means it is the first coat, or undercoat to be applied to a new (or to be patched) wall. When it is trowelled off, it is scratched with a nail to give a "key" for the top coat, or finish plaster to adhere to. Bonding plaster has incredible "stickabilty" and does not rely on an absorbent surface to bond to.

- Bonding does not need the wall underneath to be scratched or have a mechanical "key" and bonding agents, such as latex SBR adhesive are usually applied to the wall before the bonding plaster itself.

4.4.4 Coatings and Water Repellents

- Brickwork that is properly designed, constructed and maintained can be expected to satisfactorily resist water penetration under normal exposures without the application of water repellents or other external coatings. Drainage-type walls, such as brick veneer walls or cavity walls, are designed to accommodate water penetration of the exterior brickwork without damage to the interior components of the wall system through its drainage system. Use of external coatings on brick masonry should be considered only after completing repair and replacement of brick, mortar joints and other building elements, and careful consideration of the possible consequences. Although coatings are not required on properly designed, specified and constructed brick masonry, they may be used successfully to alter the appearance of a wall or to diminish the effects of certain deficiencies.

- External coatings are most effective in reducing water penetration when their intended use corresponds with the nature of the existing water penetration problem. Application of a water repellent is generally not recommended on newly constructed brick veneer walls or cavity walls. However, water repellents may be used to correct minor deficiencies that remain after completion of repairs or to reduce the amount of water absorbed by barrier walls and masonry subject to extreme exposures, such as chimneys, parapets, copings and sills. Water repellents and coatings should not replace or be considered equivalent to essential, code-required details that resist water penetration, such as flashing and weeps. Use of coatings for reasons outside their intended application rarely reduces water penetration and may lead to more serious complications with the brickwork.

4.5 REPAIR METHODS FOR MINOR AND MEDIUM CRACKS

(a) For small and medium cracks (width 0.5 mm to 5 mm) - grouting,

   For major cracks (width more than 5mm) - fixing mesh across cracks.

   (i) Cracks up to 1.5 mm in width, requiring no repair if the bricks are of absorbing type, as is usually the case in India. In case of non-absorbent bricks, rainwater is likely to penetrate through thin cracks. So it also need to be repaired thin cracks. The cracked joints should be pulled out and filled with 1 : 1 : 6 (cement, lime, sand) mortar.

   (ii) Cracks wider than 1.5 mm require repairing and method depends on the type of mortar used in the brickwork.

(b) With weak mortars, the cracks should be enlarged and crushed to a depth of about 25 mm and refilled with 1 : 2 : 9 (cement, lime, sand) mortar and repainted or re-plastered (10 cm wide strip around crack) with the same mortar.

(c) With strong mortar, bricks adjacent to the crack should be cut and new brick is replaced using 1 : 1 : 6 (cement, lime, sand) mortar. If the bricks are cracked then same procedure should be followed.

(d) Wide diagonal cracks, which are caused by the settlement of foundation, if there is a possibility of further movement, should be repaired by removing and replacing all cracked bricks. RCC stitching blocks should be used in every 5th or 6th course, that is, at about 0.5 m in the vertical direction. The stitch block in width should be equal to the thickness of the wall, in length equal to half to 2 bricks and in thickness equal to 1 or 2 bricks. It is not desirable to use mortar stronger than 1 : 1 : 6 (cement, lime, sand) for these repairs.

4.5.1 Epoxy Injection

- The injection of a low viscosity epoxy is a possible repair method for cracks. The epoxy injection to be effective, the crack must be free of dirt grease or other contaminations.

- The crack shall be between 0.02 mm and 6 mm in width. For very fine crack, less than 0.15 mm in width, the entry ports should be spaced not more than 150 mm apart.

- It is necessary to choose carefully to match the individual job requirement.

- The capability of bonding to moist masonry, shrinkage, thermal and elastic properties of hardened resins and other special needs such a fire resistance high temperature stability.

- It is relatively new work, satisfactory cleaning can often be achieved by vacuum cleaning, a head of the sealing operation. Compressed air or blasting with water or air/water mix have been suggested but the process tends to drive dust and contaminations into the bottom of cracks.

- Injection should start at the lowest post and be continued until resin appears at the next higher port. The injection nozzle is then removed from the port seal and the nozzle moved upto the next port. If the pumping pressure cannot be maintained in a cracks that appears full, the epoxy resin may leak.

- Finally when the injected resin is cured, the sealing adhesive must be removed by grinding, cutting and at port must be made good with epoxy. The cut width of about 20 mm to be filled with dry pack or epoxy mortar.

Three methods of providing entry ports :

- Drilled holes with fitting inserted and bonded in, with the adhesive used for sealing.

- Bonded flush fitting attached by means of the sealing adhesive.

- Interrupted seal using a gasket that covers the unsealed portion.

4.5.2 Grooving and Sealing

- This is the simplest and most common method of crack repair. It can be executed with relatively unskilled personnel and can be used to seal both fine pattern cracks and larger isolated cracks.

- The system can be used to repair dormant cracks that are of no structural significance, and is used to seal the cracks against the ingress of moisture, chemicals and carbon dioxide. This involves enlarging the crack along its exposed face and sealing it with crack fillers as shown in Fig. 4.18. Care should be taken to ensure that the entire crack is routed and sealed.

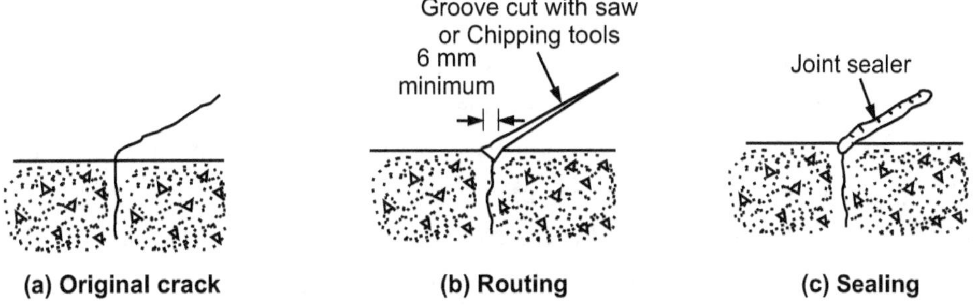

Fig. 4.18: Grooving and Sealing

4.5.3 Stitching

- In this technique, the crack is bridged with U-shaped metal units stitching dogs before being repaired with a rigid resin material. This can establish restoration of the strength and integrity of cracked section; due care is to be given to make analysis check to ensure that this will perform well under applied loads as shown Fig. 4.19. A non-shrink or an epoxy resin based adhesive should be used to anchor the legs of the dogs.

- Stitching is suitable when tensile strength must be reestablished across major cracks, although stitching will not close the crack, and it is way of stopping the movement of active crack and thereby preventing it from spreading. Stitching dogs should be of variable length and orientation and so located that the tension transmitted across the crack is not applied to a single plane within the section but us spread over an area.

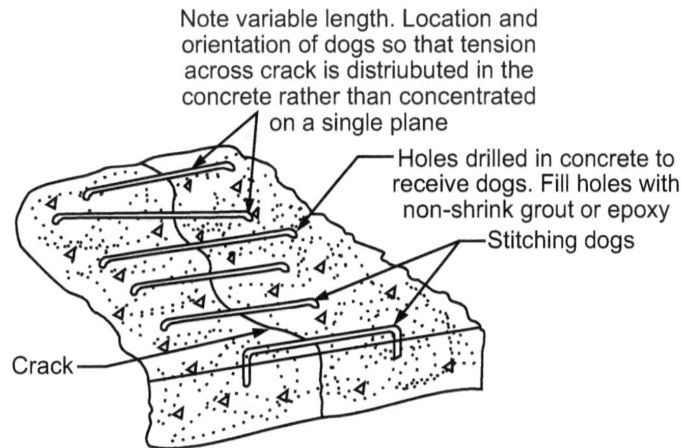

Fig. 4.19: Stitching

4.5.4 Guniting

- The process or technique that involves pumping of concrete or mortar under high pressure through a hose to a nozzle (spray-gun) and then spraying them at high velocity onto a surface either by dry-mix or by wet-mix process is called shotcreting. Usually, shotcrete is reinforced by using steel or wire mesh, steel rebars, synthetic fibres etc. Since, the dry-mix shotcreting is also called as guniting by some (yet, not by all), better to stick to the terms dry-mix shotcreting and wet-mix shotcreting rather than guniting.

Advantages of Guniting :

1.   It is extremely versatile. This process can be used to give virtually any shape to any element of a structure – curved, undulating, spheroidal and many more. This is a reason why this process is commonly adopted to add special features or shapes to swimming pools, artificial caves and waterfalls etc. It is used in sculpting work as well. It's versatility makes it an excellent candidate for a large variety of work such as tunnel lining, refractory lining, slope protection, repair works besides making it suitable for diverse uses in structures like tanks, dams, reservoirs, canals, docks, bridges, pipelines etc.

2.   Water content of the mix can be controlled instantly and at any time at the nozzle by the crew while spraying the mix onto a surface. This enables one – at any moment – to render the mix only as much dry or wet as the situation demands by controlling the addition of water at the nozzle.

3.   Dry-mix shotcrete (gunite) is quite economical, it can be quickly prepared. It is quite suitable for overhead application where wet-mix shotcrete may not stick well and may sag or fall off.

4.   If very thin linings or coatings are required to be applied on a surface or if the quantum of the work is small or in case of concrete repairs need finer treatments such as filling up of cracks or small broken patches, damage etc, guniting would be a the choice rather than wet-mix shotcreting.

5.   If the application job involves frequent stops of work for some reason or another guniting could be a much better choice over the wet-mix application.

6.   When pouring concrete (like pavement) if work is stopped and started again, those two pours would not bind together. They will be two separate pours.

- So if it is stoped at a rough edge and tried to blend the new into the old, it will create a cold joint which will look different and most likely crack there.

- When guniting. It can stop and start again without creating that "cold joint" or plane of weakness. Because of the velocity the material is applied, it will still bond together.

- The builders have more work time since they mix the cement on-site and can stop and start as needed.

- This process tends to be less expensive than shotcrete.

Disadvantages of Guniting :

- The gunite process requires a super skilled operator because that person is in charge of the sand/cement/water ratio. An error there, can ruin the quality of the concrete.

- The dry mixture could clog the hose pipe.

- Gunite produces a lot of over-spray, called rebound.

- It makes a huge mess.

- Can not be reused, so it's just wasted.

4.5.5 Shotcreting

- Shotcrete sprayed concrete is concrete or mortar conveyed through a hose and pneumatically projected at high velocity onto a surface, as a construction technique, first used in 1914. It is typically reinforced by conventional steel rods, steel mesh or fibers.

- Shotcrete is usually an all-inclusive term for both the wet-mix and dry-mix versions. In pool construction, however, shotcrete refers to wet mix and gunite to dry mix. In this context, these terms are not interchangeable.

- Shotcrete is placed and compacted/consolidated at the same time, due to the force with which it is ejected from the nozzle. It can be sprayed onto any type or shape of surface, including vertical or overhead areas.

- Masonry walls exposed to the elements and suffering from the effects of age or seismic action require to be made over by strengthening, if the cost of building a new wall after demolition of the old one is to be avoided. One of the methods that are vastly popular with engineers involved in renovation of structures is the use of shotcrete along with wire mesh reinforcement. Shotcrete reinforcement of masonry walls involves the application of concrete or mortar to surfaces by projecting the material pneumatically at very high velocity that enables placement and compaction at the same time. Besides the rehabilitation of old brick walls shotcrete is also extensively used for protecting excavated surfaces, as support for tunnel surfaces and other places where some sort of support is required.

- Shotcrete that is properly applied has very good structural properties, but if not properly done, the material becomes brittle and may turn out to be worse than the original problem it had been asked to solve. Because of its placement at high velocity, shotcrete that has been applied can be superior to concrete or mortar placed *in situ*. For the same reason, it forms a very good bond with masonry and adheres very firmly to it. Shotcrete application also completely avoids the cost of formwork that conventional concrete requires. This formwork greatly adds to the cost of concreting a structure.

Advantages of Shotcreting :

- While builders using shotcrete need to be skilled, they do not have to be as technically trained as someone working with gunite because the concrete is premixed.

- Shotcrete forms a strong and consistent coating.

- Shotcrete requires less time.

Disadvantages of Shotcreting :

- Since shotcrete is a premixed, it has to apply it quickly.

- Cracks can form from shrinkage if too much water is added to the mix.

- Shotcrete is more expensive than gunite.

- The builders might add water to the cement mixture in the cement truck to keep it from hardening. This can compromise the strength of the concrete.

Applications of Shotcrete to Old Brick Walls :

- The application of shotcrete to a brick wall for repairing or renovating a structure requires the erection of scaffolding along the surface of the wall. The extreme pneumatic pressures that are used when applying shotcrete require that this scaffolding has a stable base and is additionally anchored to the wall or by other means so that this pressure does not affect it. Shotcrete is generally delivered through concrete pumps or other grouting machines. The distance of this equipment from the point of application is of vital importance and may sometimes require portable equipment if distances or heights are too great.

- The wall to be reinforced has to be examined for its capacity to resist the high shotcrete velocity that can emanate from the nozzle. Where any doubt arises it may be advisable to support the wall temporarily during the process. This can be done by erecting a formwork behind the area that is going to be shotcreted. This formwork can be moved once the work is completed, thus reducing the total cost of the formwork.

- Wire mesh reinforcement has then to be anchored onto the wall surface, with spacers that keep the mesh slightly away from the wall surface. This can be avoided if a thin layer of mortar is applied to the wall before installation of the reinforcement. The mix to be used in the shotcrete should be of normal concrete with a larger proportion of fine aggregate and with maximum size of aggregate restricted to ¾ inch. Normal shotcrete layers may vary between 1 to 2 inches depending on the structural requirement. Because of the high velocity, shotcrete material has a rebound from the surface being treated and may lead to fairly large losses in material. On large jobs however, such shotcrete rebound material is collected and reused as aggregate.

- Once the entire surface of the brick wall has been covered, normal methods of curing and protection of newly laid concrete have to be in place. The shotcrete adds substantially to the strength of an old brick wall and can greatly increase its useful life.

4.6 REPAIRING METHODS OF MAJOR CRACKS (WIDTH MORE THAN 5MM)

- Cracks (width more than 5mm) in structures reduce their strength considerably to bear the design loads. Thus repair of such cracks is necessary to restore the designed strength of members. Repair of large cracks (cracks wider than 5mm) and crushed concrete and masonry structure cannot be done using pressure injection or grouting.

- For repair of large cracks and crushed concrete, following procedure can be adopted:

  1. The surface of cracks or crushed concrete is cleaned and all the loose materials are removed. These are then filled with quick setting cement mortar grouts.

  2. If the cracks are large, then these cracks are dressed to have a V groove at both sides of the member for easy placement of grouts.

  3. For cracks which are very large, filler materials such as stone chips can be used.

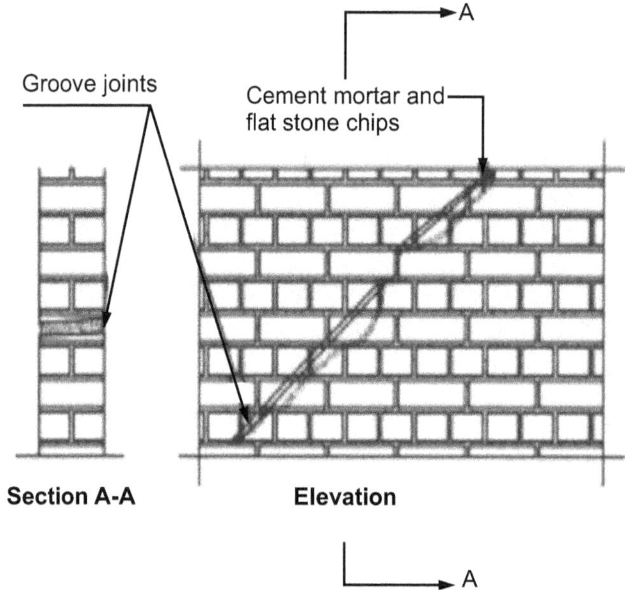

Fig. 4.20: Crack repair

  4. Additional reinforcement and shear reinforcements can be used for heavily damaged concrete members or wherever necessary based on requirements.

     These additional reinforcement should be protected from corrosion by using polymer mortar or epoxy coatings.

  5. For damaged walls and roofs, additional reinforcement in the form of mesh is used on one side or both sides of the members. These mesh should sufficiently tied with existing members.

  6. Stitching of cracks are done to prevent the widening of the existing cracks. In this case, holes of 6 to 10 mm are drilled on both sides of the crack. Then these drilled holes are cleaned, legs of stitching dogs are anchored with short legs.

     The stitching of cracks is not a method of crack repair or to gain the lost strength, this method is used to prevent the cracks from propagating and widening.

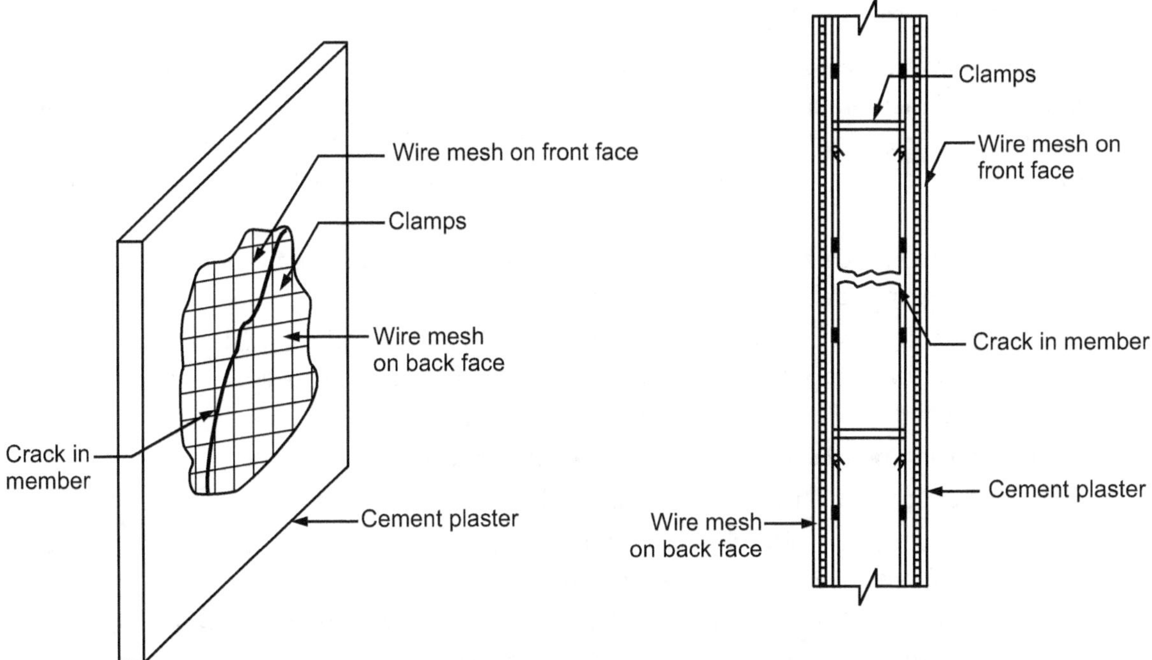

Fig. 4.21: Reinforcement meshes in repair of roof slabs and walls

4.7 DAMPNESS IN BUILDINGS

- The access and penetration of moisture content into building through its walls, floor, roof etc. is called dampness in buildings. Dampness is the presence of hygroscopic or gravitational moisture. Dampness gives rise to unhygienic conditions apart from reduction in strength of structural components of the building.

Harmful Effects of Dampness:

- Dampness causes the following harmful effects :
  1. Dampness is responsible for breeding of mosquitoes and creates unhealthy living conditions.
  2. Due to dampness, moisture travel through walls and ceilings, and creates unsightly patches and affect the aesthetics of the building.
  3. In case of lime plaster, moisture travel causes softening and crumbling of plaster.
  4. Dampness in the wall cause efflorescence and sometime dampness is responsible for disintegration of bricks, stones, tiles etc.
  5. The wall decoration (like painting) gets damaged due to dampness.
  6. The flooring get loosened because of reduction in the adhesion when moisture enters through the floor.
  7. Timber fittings (such as doors, windows, almirahs) when come in contact with damp walls or floors, gets deteriorated because of warping, buckling, dry rutting etc. of timber.
  8. Dampness in building can be very dangerous if any form of electrical fittings come in contact with it.
  9. Floor coverings are damaged. Therefore no floor coverings should be used on damp floors.
  10. Dampness promotes and accelerates growth of termites.
  11. Dampness breeds germs of dangerous diseases such as tuberculosis, neuralgia, rheumatism etc.
  12. Metal fitting get corroded and rusted due to dampness.

Methods of Preventing Dampness in Buildings :

  1. By providing DPC (Damp proof course).
  2. By surface treatment i.e. by providing damp proof paint.
  3. By integral water proofing method.
  4. By special devices i.e. by providing chajjas and by providing cavity walls etc.

4.7.1 DPC - Damp Proof Course

- It is continuous layer of impervious material applied to prevent moisture transmission. This consists in providing layers of membrane of water repellant material between the source of dampness and the part of the structure adjacent to it. This type of layer is commonly known as damp proof course (DPC) and it may comprise of materials like bituminous felts, mastic, asphalt, plastic or polythene sheets, cement concrete, etc.

- Depending upon the source of dampness, DPC may be provided horizontally or vertically in floors, walls, etc. Provision of DPC in basement is normally termed as tanking.

- Rising damp is caused by capillary action drawing moisture up through the porous elements of a building's fabric. Rising damp, and some penetrating damps, can be caused by faults to, or the absence of a damp-proof course (DPC) or damp-proof membrane (DPM).

- **General principles to be observed while laying damp proof course are :**
  - The DPC should cover full thickness of walls excluding rendering.
  - The mortar bed upon which the DPC is to be laid should be made level, even and free from projections. Uneven base is likely to cause damage to DPC.
  - When a horizontal DPC is to be continued up a vertical face a cement concrete fillet 75 mm in radius should be provided at the junction prior to the treatment.
  - Each DPC should be placed in correct relation to other DPC so as to ensure complete and continuous barrier to the passage of water from floors, walls or roof.

- **Types of DPC of Bitumen :**

  There are two types of DPC :
  - **Flexible DPC :** It is DPC when load does not crack e.g. Polythene and Bitumen.
  - **Rigid DPC :** It is DPC when loaded; it cracks e.g. Rich cement concrete 1 : 2 : 4.

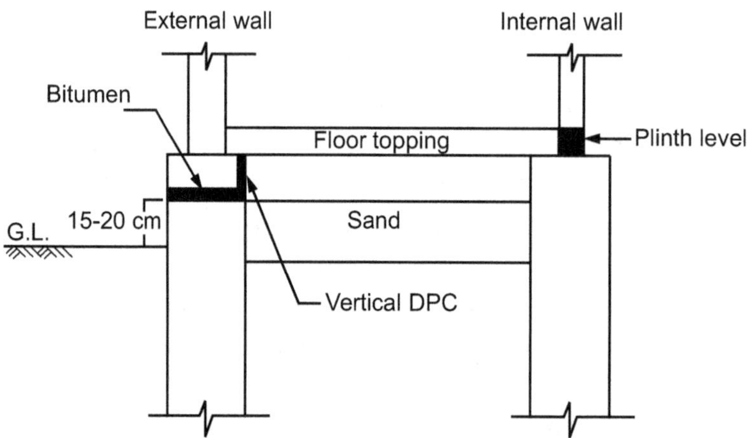

Fig. 4.22: Damp proofing

4.7.2 Bituminous Painting (Bituminous Damp Proofing)

- For internal wall we only provide horizontal DPC (175 kg/cm^2 standard pressure for bitumen). A mortar layer shall be provided before DPC. Three layers of bitumen are provided.

Three layers

1. Bitumen mastic : Bitumen mix with fine sand.
2. Bitumen felt : It is available in the form of rolled sheets.
3. Hard laid bitumen, Metal sheets e.g. Lead, copper, aluminum is provided with mortar, to avoid rusting.

4.7.3 Surface Treatment

- In the internal finish, dampness finds its way through the pores of material used in finishing. To prevent the entry of the moisture into the pores, they must be filled up. To limit the entry of moisture in surface finish a layer of water repellent substances or compounds are applied on these surfaces through which moisture enters.

- The use of waterproof metallic soaps such as calcium and aluminum stearates (Stearates are the salts and esters of stearic acid) are much effective against rainwater penetration. The finishing of outer work such as pointing and plastering surfaces must be done carefully, using waterproofing agents like sodium or potassium silicates, aluminum or zinc sulphates, barium hydroxide and magnesium sulphates etc. This method of treatment is effective only when the moisture occurring on the surface is not under pressure.

- For exposed stone or brick, wall face may be treated by spraying with water repellent solutions. For external wall plastering use of mix proportion of cement, lime, and sand mixed in 1 : 1 : 6 is found to serve the purpose of preventing dampness in the wall due to rain effectively.

4.7.4 Integral Damp Proofing

- In the integral damp proofing method certain water proofing compounds are added to the concrete mix, so that it becomes impermeable. The quantity of water proofing compound to be added to cement depends upon the manufacturer's recommendations. In general one kilogram of water proofing compound is added with one bag of cement to render the mortar or concrete water proof.

- It consists in mixing commercially available compounds in water before concrete is wet mixed. These compounds are made from chalk, talc, flutter earth or chemical compounds like calcium chloride, aluminum sulphate, calcium chloride etc

4.8 FOUNDATION SETTLEMENT

4.8.1 Causes of Foundation Settlement

- The causes of foundation settlement are rarely due to the design (or under-design) of the structure itself. More commonly, damage is caused as changes occur within the foundation soils that surround and support the structure. The following briefly describes a few of the more common causes of foundation settlement :

(i) Weak Bearing Soils :

- Some soils are simply not capable of supporting the weight or bearing pressure exerted by a building's foundation. As a result, the footings press or sink into the soft soils, similar in theory to how a person standing in the mud sinks into soft, wet clay.

- In such cases, footings may be designed to spread the load over the weak soils, thereby reducing potential foundation settlement. However, the majority of settlement problems caused by weak bearing soils occur in residential construction, where the footings are designed based upon general guidelines and not site-specific soil information. e.g. Leaning tower of pisa, itly.

(ii) Poor Compaction :

- Placement of fill soils is common practice in the development of both commercial and residential subdivisions.

- In general, before a foundation can be constructed on a plot, hilltops are cut down and valleys are filled in order to create buildable lots. Properly placed and compacted fill soils can provide adequate support for foundations, and are sometimes brought in from off-site locations.

- When fill soils are not adequately compacted, they can compress under a foundation load resulting in settlement of the structure.

(iii) Changes in Moisture Content :

- Extreme changes in moisture content within foundation soils can result in damaging settlement. Excess moisture can saturate foundation soils, which often leads to softening or weakening of clays and silts. The reduced ability of the soil to support the load results in foundation settlement. Increased moisture within foundation soils is often a consequence of poor surface drainage around the structure, leaks in water lines or plumbing, or a raised groundwater table.

- Soils with high clay contents also have a tendency to shrink with loss of moisture. As clay soils dry out, they shrink or contract, resulting in a general decrease in soil volume.
- Therefore, settlement damage is often observed in a structure supported on dried-out soil. Drying of foundation soils is commonly caused by extensive drought-like conditions, maturing trees and vegetation and leaking sub-floor heating, ventilation, and air conditioning (HVAC) systems.

(iv) Maturing Trees and Vegetation :

- Maturing trees, bushes and other vegetation in close proximity to a home or building are a common cause of settlement. As trees and other vegetation mature, their demand for water also grows.
- The root systems continually expand and can draw moisture from the soil beneath the foundation. Again, clay-rich soils shrink as they lose moisture, resulting in settlement of overlying structures. Many home and building owners often state that they did not have a settlement problem until decades after the structure was built.
- This time frame coincides with the maturation and growth of the trees and vegetation.
- Foundations closer to the surface are more often affected by soil dehydration due to tree roots than are deep, basement level foundations. As a general rule, the diameter of a tree's root system is at least as large as the tree's canopy.

4.8.2 Foundation Underpinning and Piering

- There are two methods used to support sinking foundations :

1. Foundation Push Piers :

- The Foundation Support works Push Pier System, permanently stabilizes foundation and offers the best opportunity to lift building back to level without the expense and disruption of a full foundation replacement. If cracks are detected, uneven floors, jamming doors or windows that stick, then the signs of the settling foundation may be witnessed.
- Foundation Support works Push Pier system, utilize high-strength steel pier sections that are hydraulically driven through heavy-duty steel foundation brackets to reach deep down to competent load-bearing strata. The piers have the ability to reach far below the problem soils and do not rely on friction for capacity. Used in both residential and commercial applications, pier installation can be completed year-round without the major disruption of other methods. Foundation Support works Push Piers effectively stabilize settling foundations and provide the best opportunity to lift building back to a level position.

2. Soil Improvement Techniques :

- The presence of soil unsuitable for supporting structures in construction sites, lack of space and economic motivation are primary main reasons for using soil improvement techniques with poor sub-grade soil conditions rather than deep foundation.
- Several methods are commonly used to reduce the post construction settlement, increase the shear strength of the soil system, increase the bearing capacity of the soil, and embankments. There are many available improvement techniques that can be used for the purposes of improving bearing capacity, increasing shear strength and decreasing consolidation settlement of saturated medium clay such as soil replacement, preloading with vertical drains, stone columns, stabilization with additives and thermal methods.

(a) Pre compression or preloading :

- It is simply a process to place a surcharge fill on the top of the soil that requires large consolidation settlement to take place before construction of the structure. Once sufficient consolidation has occurred, the fill can be removed and construction process takes place. In general, this technique is adequate and most effective in clayey soil. Since clayey soils have low permeability, the desired consolidation takes very long time to occur, even with very high surcharge load. Therefore, with tight construction schedules, preloading may not be a feasible solution. Hence, sand or vertical drains may be used to accelerate consolidation process by reducing the drainage paths length.

(b) Sand Compaction Piles :

- Sand compaction piles are one of the potential methods for improving ground stability, preventing liquefaction, reducing settlement and similar applications. This method involves driving a hollow steel pipe into the ground. The bottom is closed with a collapsible plate down to the required depth and then pipe is filled with sand. The pipe is withdrawn while the air pressure is directed against the sand inside it. The bottom plate opens during withdrawal and the sand backfills the voids created earlier during the driving of the pipe. The sand backfill prevents the soil surrounding the compaction pipe from collapsing as the pipe is withdrawn. During this process, the soil gets densified.

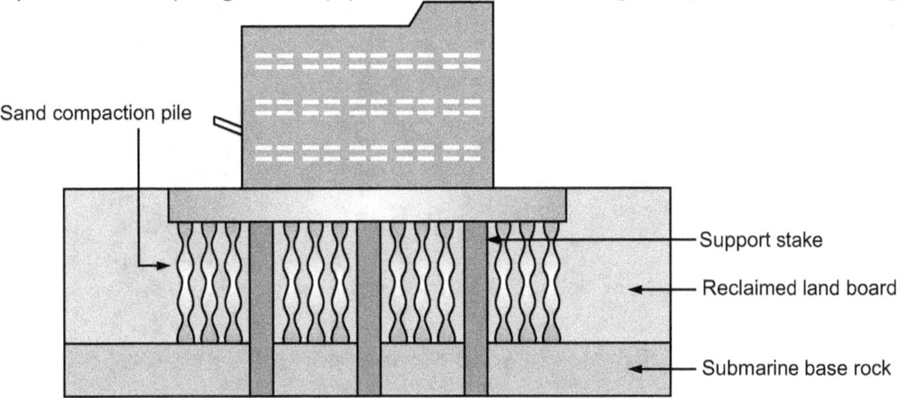

Fig. 4.23 : Sand compaction piles

Design of sand compaction piles depends on :

- Stability of the sand column.
- Piles and soil which are subjected to equal vertical deformation.

Advantages of sand compaction piles :

- Sand used is cheaper as compared to other ground improvement techniques so it is economical.
- Construction of sand column is extremely fast so required time is less.
- After creating the hole, it is fully supported by casing during construction that prevents the possibility of collapse.

Disadvantages of Sand Compaction Piles :

- Sand compaction piles have low stiffness.
- Larger percentage of replacement of weak soil is required.
- These piles do not have high permeability to function as effective vertical drains during earthquakes
- Sand compaction piles are more likely susceptible to liquefaction during earthquake.

Types of Sand Compaction Piles :

- There are two types of Sand compaction piles method based on the system deployed. First one is vibratory system with vibro-hammer and the other one has a non-vibratory system with forced lifting or driving device.

(i) Vibratory Sand Compaction Piles :

- The vibratory sand compaction piles were developed 50 years ago and has been used in more than 380,000 km of improved ground. But the vibro-hammer used in this method has a negative effect in the form of vibration and noise to the surrounding environment which makes it difficult to use this method in the urban areas or at locations close to existing structures.

(ii) Non-vibratory Sand Compaction Piles:

- To avoid these problems, a system with a non-vibratory sand compaction pile method was developed, which does not require impact or vibration on the driving device to penetrate into the ground. The equipment consists mainly of a sand compaction pile driving device used as a base machine and a forced lifting or driving device with a rotary drive motor to rotate the casing pipe.

Procedure for Non-vibratory Sand Compaction Piles:

- Set the casing pipe to the predetermined place.
- By operating the forced lifting or driving device, install the casing pipe into the ground while rotating.

- The sand is fed through upper hopper after the pipe has reached the required depth.
- Casing pipe is pulled up which results in the sand being pressed out to the void by compressed air.
- The casing is then extracted along with the compaction of the pressed out sand pile so as to enlarge it.
- Above procedure is repeated till the sand piles are formed to the ground surface.

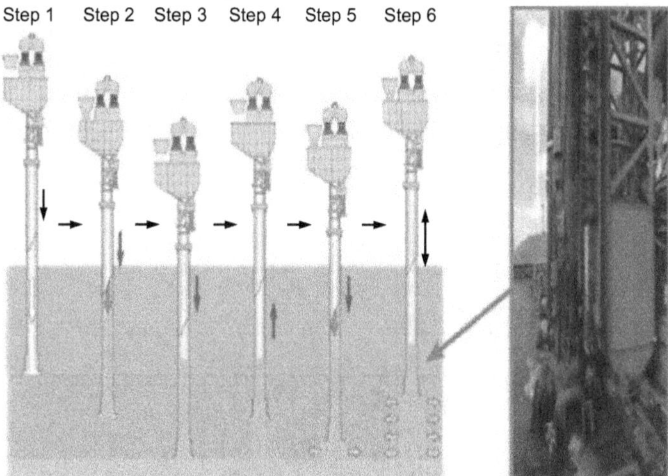

Fig. 4.24 : Process of sand compaction piles method

(c) Stone Column :

- It is a ground improvement technique used to improve the load bearing capacity and reduce the soil settlement. It is also called as granular columns or granular piles. This technique is also known as vibro replacement. In this technique, dense aggregate column (stone columns) is constructed by means of a crane-suspended down hole vibrator.

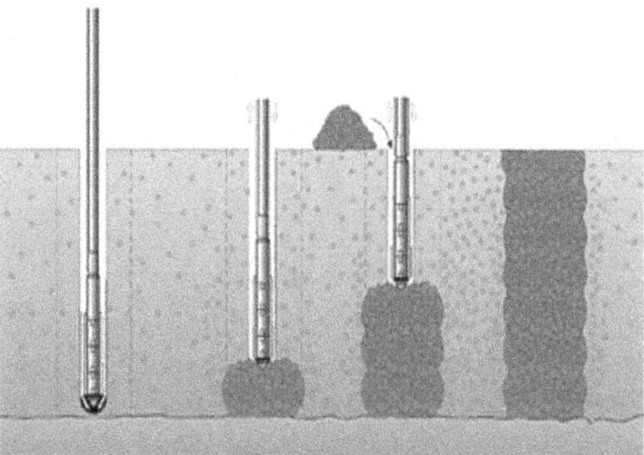

Fig. 4.25 : Stone column

Function :

- Installation of stone column improves ground by reducing soil settlement. Due to its higher modulus of elasticity than soil, it absorbs more load than soil and reduces overall settlement. Since applied load distributes in between soil and stone column in the ratio of their stiffness, the bearing capacity of soil also increases. Stone aggregates are used to fill stone column. Water can easily pass into the stone column. So, stone column helps in excess pore water pressure mitigation and accelerates the consolidation process.

Construction of Stone Column :

- Stone columns are constructed by experienced contractors using specialist equipment. The construction uses an excavator with a vibrating probe to feed stone into the ground, creating a vertical column of stone. Some stone column rigs feed stone into the ground through the vibrating probe, exiting at the bottom, and other rigs require the stone to be fed in from the ground surface down the vertical hole in the ground. Both types use a vibrating probe that densies the surrounding soils to help feed the stone into the ground.

Effect of Stone Column on Ground :
- Stone columns help to limit the amount and consequences of future liquefaction by :
  - Densifying the soil through vibration and introducing stone into the soil.
  - Reinforcing the soil forming a stiff composite soil mass. By achieving this, the non-liquefying soil crust is thickened and stiffened to reduce the likelihood of undulations, tilt and uneven ground surface subsidence from liquefaction of the underlying soil layers, therefore reducing damage to the house foundations.
  - In addition, stone columns may sometimes provide the soil with an increased drainage path to help reduce excess pore water pressure that can lead to liquefaction, so the columns can reduce the consequences of liquefaction when this occurs.

(d) Grouting Cement Slurry :
- Cement has been the oldest binding agent since the invention of soil stabilization technology in the 1960s. It is commonly used to stabilize a wide range of soils, adding enough volume, increasing soil content and making the soil harder and harder to work with. This may be due to the use of cement to stabilize the wider soils. In this manner, the cement is mixed with water and soil by special equipment on site. Physical and chemical reactions occur in cement and clay, cement alignment attaches the clay to the soil, but does not impede the structure of the soil. Hardening process is affected by curing the soils, chemical and physicochemical properties, water-cement ratio and temperature and compression level. On the other hand, the nature and type of treated soil.
- The amount of cement consumed, the placement and the preventive conditions affect the determination of the optimal ratio of soil-cement.

Important Points
- The principle mechanism causing crack in structures are Moisture change, Thermal movement, Elastic deformation, Creep, Chemical reaction, Foundation movement and settlement of soil, Growth of vegetation.
- In load bearing structures, where a roof slab undergoes alternate expansion and contraction due to temperature variation, horizontal cracks may occur (shear cracks) in cross walls, due to inadequate thermal insulation or protective cover on the roof slab.
- A frequency of thermal cracks occurring in buildings is the formation of a horizontal crack in support of a brick parapet wall or brick-less-iron railing on an RCC cantilever balcony.
- Horizontal cracks are caused by deflection of the slab and lifting up the slab edge, combined with horizontal motion in the slab due to shrinkage. These cracks appear a few months after construction and are more prominent if the span is large. Due to the light vertical load on the wall, these cracks are confined to the topmost floor, due to which, the end of the slab is encountered without much restraint. In the lower storeys, vertical loads of the upper storey prevent the corners from lifting.
- Diagonal cracks are caused by differential stresses in the internal and external load bearing walls, thereby binding the cross walls. When the walls are unevenly filled with wide variations in stress at different parts, excessive shear stress develops which causes cracks in the walls.
- Vertical cracks are caused by shrinkage upon initial drying of the RCC roof slab, as well as thermal contraction, which causes inward pull on the walls in both directions. These cracks can be prevented by providing the proper movement joint i.e. the joint between the slab and the supporting walls.
- Vertical cracks around the stair case are caused by drying shrinkage/elastic shortening and thermal movement in the building. Generally, these cracks are not very obvious, and can be reduced by delaying the rendering/plastering to cause shrinkage/elastic deformation of the masonry/concrete.
- Vertical cracks in the side walls are mainly due to thermal expansion, which increases with moisture expansion of the brickwork and is noticeable during hot weather. Such cracks occur more in a building built in cold weather. These cracks begin at the DPC level and go up to the ward, and are more or less straight and pass through masonry units.

- Grouting is a process of filling the cracks, voids under pressure in masonry structural members for repairing of cracks, strengthening of damaged masonry structural members. Grout is a flowable plastic material and should have negligible shrinkage to fill the gap or voids completely and should remain stable without cracking, de-lamination or crumbling.

- The process or technique that involves pumping of concrete or mortar under high pressure through a hose to a nozzle (spray-gun) and then spraying them at high velocity onto a surface either by dry-mix or by wet-mix process is called shotcreting.

- Shotcrete sprayed concrete is concrete or mortar conveyed through a hose and pneumatically projected at high velocity onto a surface, as a construction.

- Efflorescence consists of white deposits on the brick surface left when moisture carrying dissolved salts evaporates.

- The movement of moisture to a wall through the brickwork by capillary action is referred to as rising damp. This condition appears as a rising water line or "tide mark" on the wall caused by the soluble salts in the groundwater. Consequently to reduce the potential for rising damp, a damp proof course should be installed.

- Damp proof course is continuous layer of impervious material applied to prevent moisture transmission. A common example is polyethylene sheeting laid under a concrete slab to prevent the concrete from gaining moisture through capillary action. A DPM may be used for the DPC.

- Rising damp is caused by capillary action drawing moisture up through the porous elements of a building's fabric. Rising damp, and some penetrating damp, can be caused by faults to, or the absence of a damp-proof course (DPC) or damp-proof membrane (DPM).

- Membrane Damp Proofing consists in providing layers of membrane of water repellent material between the source of dampness and the part of the structure adjacent to it. This type of layer is commonly known as damp proof course (DPC) and it may comprise of materials like bituminous felts, mastic, asphalt, plastic or polythene sheets, cement concrete etc.

- Depending upon the source of dampness, DPC may be provided horizontally or vertically in floors, walls, etc. Provision of DPC in basement is normally termed as tanking.

- The causes of foundation settlement are rarely due to the design (or under-design) of the structure itself. More commonly, damage is caused as changes occur within the foundation soils that surround and support the structure.

Questions for Practice

1. Explain the causes of the cracks.
2. Explain the remedies to the cracks.
3. Explain types of cracks.
4. Explain briefly the cracks in machinery structure ?
5. Explain the Damping effect and its repair technique.
6. Enlist the repairing methods for major cracks and explain any one briefly.
7. Explain the various causes for wall cracks and their probable location.
8. Explain the repair technique for damages in the civil structure with justification.
9. Explain the crack in : (a) RCC frame structure (b) Free standing wall.
10. Give objectives of cracks repairs?.
11. Explain the following terms : (1) Shortcreting, (2) Guniting.
12. Explain the various methods improving the bearing capacity of foundation.
13. Enlist the causes and remedies of foundation settlement.
14. Explain improvement technique of foundation settlement.
15. Write advantages and disadvantages of sand compaction.

. . .

Chapter **5**

MAINTENANCE AND REPAIR METHODS FOR RCC CONSTRUCTION

Weightage of Marks = 14, Teaching Hours = 14

Syllabus

5.1 Probable location of cracks in RCC elements, various causes of RCC failure.

5.2 Causes of dampness in roof slab and its repair techniques such as mud phuska with brick tile topping, lime concrete terracing, ferro-cement topping and brick coba.

5.3 Repair methods for cracks in RCC structures such as epoxy injection, grooving and sealing, stitching, rebaring, grouting, spalling replacement, jacketing, shotcrete and gunitting.

5.4 Repair of corroded RCC element : exposing and undercutting rebar, cleaning reinforcing steel, compensating reinforcement and protective coatings.

5.5 Repair methods for honeycomb and larger voids.

Objectives

After learning this chapter, student will be able to

- Explain the probable crack location in RCC and causes of RCC failure.
- Explain the causes of dampness in roof slab and its various repair techniques.
- Know the repair methods for the cracked RCC elements.
- Explain the relevant repair methods for cracks in RCC structures.
- Know the repair of corroded RCC elements, honeycomb and large voids in the given structure.

INTRODUCTION

- Concrete is the most widely used and versatile construction material possessing several advantages over steel and other construction materials. However very often one come across with some defects in concrete. The technique to be adopted for repair or restoration of the structure depends on the cause, extent and nature of damage, the function and importance of the structure, availability of suitable materials and facilities for carrying out repair, and a thorough knowledge of the long-term behavior of the materials used for the repair work.

- Depending upon the requirement, the repairing technique may be of a superficial (cosmetic) nature or, in some cases, may involve the replacement of part or whole of the structure.

5.1 CRACKS IN RCC ELEMENTS

- Concrete is a composite material that consists of embedded particles, aggregates with a binding medium. In cement concrete, the binding medium is the mixture of cement and water. The concrete is said to be durable when changes occur at a rate, which does not adversely affect its performance within its intended life.

- Reinforced Cement Concrete is a composite structural material. It is used for various structural uses. But RCC has not proved to be durable due to various factors, mainly variations in production, loading conditions in service life and subsequent attack of the environmental factors. If RCC is well-mixed, appropriately compacted and cured it is considerably water tight and durable. It continues to be in the condition as long as capillary pores and micro-cracks in the interior do not form interconnected pathways.

(5.1)

- Many reinforced concrete structures, within a period of 15-20 years deteriorate considerably. The cracking and spalling of concrete, mainly involve corrosion of reinforcement. In many of the cases, penetration of water and aggressive chemicals during the service life of structures, is the primary reason for the problem. The natural causes for deterioration are carbonation, chloride ingress, sulphate attack, alkali-silica reaction, leaching and freeze-thaw. The first line of defense against any of the deterioration process is impermeability of concrete.

- Although it is difficult to generalize the causes of deterioration due to interacting nature of various factors, efforts have been made to group the various types as physical and chemical. The micro-structure of concrete material continuously changes its response to penetration of water, CO_2, oxygen. This is also influenced by local conditions of temperature, humidity and pressure.

- Following are the two stages of deterioration of concrete :

  1. The voids and micro-cracks in the interfacial zone are inter-linked due to weathering effects and loading. When the inter-linked network of micro-cracks between the cement paste and coarse aggregate or reinforcing steel gets connected to any other crack present at the concrete surface, this provides the passage of fluid into the interior of concrete.

  2. The amount of penetration in concrete increases considerably. Water, oxygen, carbon dioxide and acidic ions etc. are able to enter easily into concrete. The existence of these elements allows various physical-chemical interactions. As a result of which, the material ultimately undergoes cracking, spalling and loss of mass which results in loss of strength and stiffness.

Capillary porosity	High w/c ratio, Inadequate curing.
Air voids	Improper compaction.
Micro cracks	Loading effects, Weathering, Secondary effects.
Macro cracks	Placement, Hardening process, Intrinsic chemical attack, Corrosion of reinforcement.

5.1.1 Location of Cracks

- Several types of cracks occur in RCC elements due to shear stress, corrosion of reinforcement, insufficient rebar cover, bending stress and compression failure etc.

Cracks in beams/slab due to increased shear stress :

- Cracks appear in RCC beams/slab near the support such as wall or column due to increase in shear stress. These cracks are called as shear crack and are inclined inwards at 45° with the horizontal. These cracks near beam supports can be avoided by providing additional shear reinforcements such as bent up bars or stirrups near the support where the shear stress is maximum. Shear stress is maximum at a distance of effective depth/2 from the support.

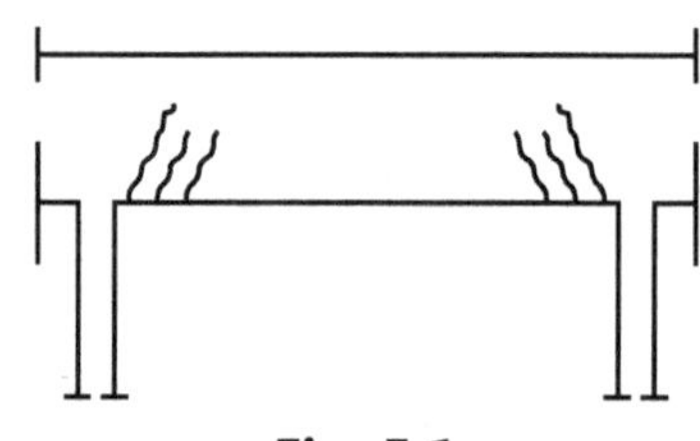

Fig. 5.1

Cracks in RCC members due to insufficient concrete cover or corrosion :

- Bottom of the beam and slab, sides of column are exposed to environment. If the cover to reinforcement is insufficient, then reinforcement is likely to corrode and cracks due to corrosion of reinforcement appear on the member. These cracks commonly appear near the reinforcement along its length. Cracks due to reinforcement corrosion may cause spalling of concrete in severe cases. It can be prevented by maintaining good quality control during its construction like, provision of adequate rebar cover as per environmental conditions.

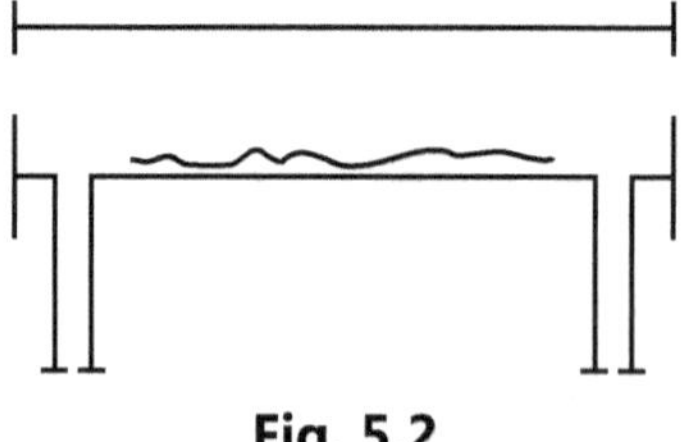

Fig. 5.2

Cracks parallel to main steel in case of corrosion :

- These cracks also appear due to corrosion of reinforcement but these appear parallel to main reinforcements. The cause of this corrosion is also due to provision of insufficient reinforcement cover which leads to corrosion of main reinforcement.

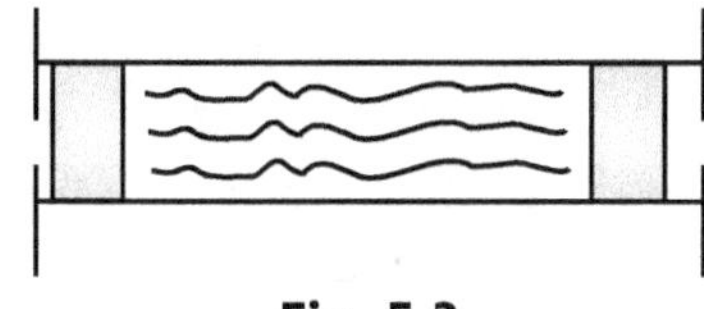

Fig. 5.3

Cracks due to increased bending stress in beams :

- Cracks due to increased bending stress in beams appear near the center of span of the beam at an angle of 45° with horizontal and at centre of bottom in slab as the bending moment is maximum at that point. If the reinforcement provided is insufficient to carry the load, bending stress increases which leads to increased deflection at the middle span. Cracks due to increased bending moment can be prevented by providing adequate main reinforcement at the mid-span. Care should be taken during design to consider all the probable loads and load combinations for its design.

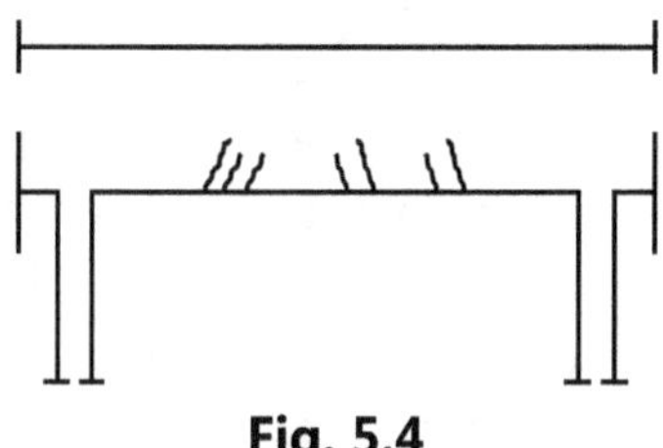

Fig. 5.4

Concrete cracks caused by premature drying :

- There are two common types of cracks due to premature drying.
    - (ii) Crazing cracks are very fine surface cracks that look like spider web or shattered glass. When the top of a concrete slab loses moisture too early, crazing cracks are likely to appear. These crazing cracks are not of structural concern.
    - (ii) Crusting cracks typically occur during the concrete stamping process. It adds texture or pattern to concrete surfaces. On sunny or windy days where the top of the slab dries out quicker than the bottom, the top of the concrete surface can become crusty. When the stamp is embedded, it pulls the surface apart near the stamped joints and causes small cracks around the outside edges of the "stones". Proper site preparation, a quality mix, and good concrete finishing practices can go a long way towards minimizing the appearance of cracks and producing a more aesthetically pleasing concrete project.

5.1.2 Physical Deficiency

- **High w/c ratio and Inadequate curing :** High capillary porosity in cement paste allows the aggressive chemicals to penetrate easily and allows the concrete/reinforcement to get affected at an accelerated rate.
- **Poorly graded aggregates :** Porous concrete due to air voids and allows the aggressive chemicals to penetrate easily and allows the concrete/reinforcement to get affected an accelerated rate. This leads to corrosion of reinforcement.
- **Inadequate compaction :** Porous concrete due to air voids and allows the aggressive chemicals from its environment to penetrate easily and allows the concrete/ reinforcement to get affected at an accelerated rate and initiate the onset of corrosion.
- **Shuttering joints not slurry tight :** Honey combed concrete due to bleeding when cement paste is replaced by air voids near surface and will easily allow penetration of aggressive chemicals.
- **Cover thickness being lesser :** Protective cover thickness against external/environmental chemical attack reduced and allows the concrete/reinforcement to get affected early.
- **Wrong placement of reinforcement :** Wrong placement of reinforcement may lead to structural or bond failures.
- **Heating/Cooling :** Surface disintegration and micro-cracking allows cracking leading to exposure of reinforcement to climate.
- **Wetting/Drying :** Increase in capillary porosity due to leaching away of water soluble salts results in depletion of water soluble calcium hydroxide reducing the alkalinity.

5.1.3 Chemical Deficiency

- Carbonation spreads to reinforcement level and depassivate reinforcement and initiate corrosion.
- Volume of outer layer increases and pressure is exerted by this increased volume of corrosion product on adjacent concrete.
- Cracking of concrete due to this pressure allows easy permeation of atmospheric gases.
- The process of corrosion accelerates and causes increased cracking and spalling of concrete.
- cross-section of steel bar reduces due to corrosion.
- Corrosion product behind reinforcement pushes bar outward to make it buckle and fail in compression to cause collapse.
- **Chloride attack :** Steel reinforcement depassivates locally and galvanic cells are formed. This initiate corrosion of bar. The chloride ion acts as current carrier in presence of water and causes localized corrosion of reinforcement.
- **Sulphate attack :** When sulphates react with Calcium Aluminate Hydrate (C-A-H) in cement paste, it forms expansive compound. This compound exerts bursting pressure and causes disintegration and cracking of concrete up to depth of permeation. This allow the aggressive chemicals from environment to penetrate easily. The reinforcement may get affected at an accelerated rate and initiate the onset of corrosion.

Table 5.1 : Intrinsic Cracks in Concrete

Type of intrinsic cracking (not caused by structural loading)	Sub-division	Most common location of occurrence	Primary cause (exluding restraint)	Secondary causes/ Factors	Remedy, assuming basic redesign is impossible (in all cases reduce restraint)	Time of appearance
Plastic settlement cracks	Over the reinforcement	Deep sections.		Rapid early drying conditions.	Reduce bleeding do air entrainment or revibrates mildly.	Ten minutes to three hours
	Arching	Top of columns trough and waffle slabs.	Excess bleeding.			
	Change of depth					
Plastic shrinkage cracks	Diagonal (may be normal to wind direction)	Read and slabs.	Rapid early drying	Low rate of bleeding and fast surface evaporation.	Improve early curing and towel.	Thirty minutes to six hours.
	Random	Reinforced concrete slabs.	Ditto plus steel near surface.			
	Over reinforcement (even mesh type)	Reinforced concrete slab.				
Early thermal construction cracks	External restraint.	Thick walls.	Excess heat generation.	Rapid cooling, curing by relatively cold water.	Reduce heat and/or insulate.	One day to two or three weeks.
	Internal restraint.	Thick slabs.	Excess temperature gradients.			

Long-term drying shrinkage cracks		Thin slabs (and walls).	Absence of movements joints or inefficient joints.	Excess shrinkage and inefficient curing.	Reduce water content and improve curing.	Several weeks or months.
Crazing cracks (occur only on surface)	Against forwmwork. Floated concrete.	Fair faced concrete slabs.	Impermeable formwork. Over trowelling.	Rick mixes. Poor curing.	Improve curing and finishing.	One to seven days sometimes such later.
Cracks due to corrosion of reinforcement (expansive reaction can lead to spalling of concrete)	Natural and slow, or fast if excessive Calcium chloride present.	Columns and beams Precast concrete	Lack of cover and dampness. Excess calcium chloride and dampness.	Poor quality concrete.	See details ahead.	More than about two years.
Cracks due to alkali-aggregate reaction (expansive reaction)		(Damp locations)	Reactive silicates and carbonates in aggregates acting on alkali in cement.		See details ahead.	More than five years.

5.2 DAMPNESS IN ROOF SLAB

- **Causes of Dampness in Roof Slab :**
  1. When the roof slab is exposed to heavy showers of rain it becomes a major source of leakage in a structure particularly when it is not suitably protected. Leakages from roof slab and joints permit rainwater to enter in a structure.
  2. The moisture is deposited on the ceilings etc. due to condensation process.
  3. When the structure is located on a site, which cannot be easily drained off, the possibility of leakage increases. This leakage will enter the structure.
  4. The orientation of a building is also an important factor. The face gaining less sunshine and heavy showers of rain may become damp and leaky.
  5. For flat roof, the slope may lead to storage rain water on slab. This water may further penetrate in the structure.
  6. Bad workmanship in construction such as improper waterproofing, defective joints in the roofs, defective rain water drain and water supply pipe connections, improper connection of the walls etc. may also contribute in dampness.

5.2.1 Leakage Through Roofs

- The causes of seepage/leakage through roofs are as under :

Cause	Remedy
(i) Lack of proper slope causing stagnation of water	Adequate slope should be provided to prevent stagnation of water.
(ii) Improper drainage system	Sufficient drainage pipes should be provided.
(iii) Lack of drip mould, coping etc.	Drip mould at junction point and coping on the top of the wall should be provided.
(iv) Leakages through water pipe connections and joints	Maintenance of water supply pipe connections and fitting should be leakproof.
(v) Poor quality of construction	Good workmanship should be adopted during construction.

5.2.2 Effect of Leakage

- The prominent effect of dampness is as follows :
  (i)   Unhealthy conditions are experienced by the tenants in a damp building.
  (ii) The steel used in the construction of the building is corroded which further damages the surrounding material.
  (iii) Unpleasant patches are observed on the wall surfaces and ceilings.
  (iv) In a damp or leaky atmosphere decay of timber takes place rapidly.
  (v)   The electric fittings are damaged due to dampness and may also result in short circuit.
  (vi) The materials used as floor coverings are also damaged.

5.2.3 Measures to Avoid Seepage/Leakage in Buildings

- Before every monsoon season, a thorough inspection of the roof should be carried out and any crack/damages to top of parapet, coping, plaster, pointing should be repaired. To prevent the need of costly repairs such annual/regular repairs shall be carried properly.

- Expansion joints can be perennial source of leakage if they are not provided and maintained in satisfactory condition. For preventing leakage at expansion joints as well as their smoother functioning for expansion. The sealing compound, which is used for sealing the expansion joints, should be inspected once a year carefully about its integrity. If the condition of the sealing at joint shows signs of decay, it should be replaced. If replacement is done in time, there will be huge saving in maintenance of buildings.

- It is essential to provide adequate slope to the flat and sloping roof to ensure effective drainage. The slope of roof should be such that the water does not accumulate and drained off quickly under gravity. A slope of 1 in 100 or steeper, depending upon the type of water proofing system, is required to drain off water effectively.

- Adequate rain water pipes shall be provided in suitable numbers and size, to allow the effective drainage of water. This depends upon the area of roof and intensity of rainfall of the region in which building is located. Average rainfall intensity may be obtained from local Indian meteorological department office. For better drainage effect, rain water pipes shall have bell mouth inlet at the roof surface. The spacing in between outlet pipes shall be less than 6 m.

- Junction of roof slab with parapet wall is a susceptible location for leakage. It is necessary that proper care shall be taken for detailing at the junction of roof and vertical face of parapet wall. The fillet of 75 mm should be provided all along the junction of parapet wall with roof. Sloping coping on the top of the parapet wall shall also be provided along with the provision of drip course on either side. Water proofing system should be stretched from roof to parapet wall for a height of minimum 150 mm with a chase.

5.2.4 Various Water Proofing System

  (i)   Lime concrete terracing.
  (ii) Water proofing with Mud-Phuska treatment.
  (iii) Water proofing with polyethylene films.
  (iv) Water proofing with polymer cementitious slurry coatings.
  (v)   Waterproofing with polymer modified bitumen membranes.
  (vi) Waterproofing using bitumen felts.

5.2.4.1 Mud-Phuska Treatment

- It is an insulating cum water-proofing treatment used in hot and dry region, where rainfall is moderate. It consist of the following layers :
  1. A coat of hot bitumen 80/100 or equivalent.
  2. A layer of mud phuska with average thickness of 100 mm consisting of puddled clay conforming to IS : 2115.

3. A layer of 25 mm mud plaster. The plaster consists of puddled clay mixed with chopped straw 30 to 35 kg per cubic meter of soil.

4. 1-2 layer of tile laid on a bed of mud mortar. Pointing in 1 : 3 cement sand mortar is necessary.

(A) Material :

- **Soil for Mud Phuska :** It shall be free from coarse sand, vegetable matter, harmful and efflorescent salts and gravel. Usually soil suitable for making clay bricks is suitable for mud-phuska.

- **Mud Plaster :** It shall be prepared from soil conforming to above. The dry soil shall be reduced to fine powder and mixed with water in a pit, adding wheat straw 6% by mass and cow dung 12% by mass. The mixture shall be allowed to rot for a period of not less than 7 days. The mixture shall be a homogeneous mass. It shall be free from lumps and clods. Slump test may be carried out as per the procedure laid down in IS : 1199. The slump should be about 70 mm.

- **Mud Mortar :** It is used as bedding under brick tile layer. It shall be prepared in the same manner as mud plaster without any addition of fibrous reinforcing material and binding material. The mud mortar shall be used immediately without any rotting period.

- **Brick Tiles :** These shall conform to the requirements given in IS : 2690 (Part-1) or IS : 2690 (Part-II).

(B) Application Procedure :

Various steps involved in this system are as under :

- **Preparatory works :** Prior to application of treatment, preparatory works shall be completed. It consist of filling of cracks by cement sand slurry, provision of adequate number of drain outlets, provision of 75 mm fillet at junction of roof slab with parapet wall, provision of a groove/chase in parapet wall etc. The details of the same are described in IS : 3067 - Code of Practice for General Design Details and Preparatory Works for Damp Proofing and Water Proofing of Buildings.

- **Cleaning of roof surface :** The surface of roof, the parapet, gutters, drain mouths etc. shall be thoroughly cleaned of all foreign matter. As the water proofing treatment is to be applied over this surface it has to be cleaned off fungus and dust etc. by wire brushing and dusting.

- **Application of bitumen over cleaned surface :** After cleaning of roof surface, a coat of hot bitumen (residual type bitumens 80/100) over the roof surface shall be applied evenly over the entire surface without any gap. Bitumen coat is extended over the vertical surfaces meeting with the slab. Bitumen commonly used is residual type petroleum bitumen of grade 80/100 or hot cutback bitumen. Residual type bitumen is heated to a temperature of not less than 165°C and not more than 170°C. The quantity of bitumen to be spread per 10 sq. m. of the surface is approx. 17 kg.

(C) Laying of Mud-Phuska :

- **Preparation :** The soil shall be stacked in required quantities in about 300 mm high stacks over a level ground and the top surface divided into suitable compartments of convenient size by bunding. The estimated quantity of water corresponding to optimum moisture content shall be added about 12 hours before the use and allowed to soak. The stacks of soil shall then be worked up with spades and hands to ensure proper mixing at the time the using soil.

- **Laying :** The Mud-Phuska shall be carried to the surface to be covered and laid in loose thickness not greater than 150 mm. The surface shall then be leveled to the slope of 1 in 40. The surface shall then be rammed with wooden rammers and 'thappies' to obtain maximum density. Generally, a Mud-Phuska layer, laid to a compacted thickness of 100 mm or more. The surface shall be allowed to dry for 24 hours. The cracks shall be filled with a grout of the binder material, if any.

- **Applying mud plaster :** Mud plaster shall be laid to a total thickness of 25 mm over the surface, After laying the mud phuska. The plaster may be applied in one or two coats of 15 mm and 10 mm. After the application, the coat of plaster shall be allowed to dry. The surface shall be checked again for slope and evenness with a straight edge and spirit level. Wherever necessary it shall be made up by application of the plaster.

- **Paving with brick tiles :** The brick tiles shall be laid flat on a thin layer of mud mortar. After application of mud plaster. The tiles shall be laid close to each other such that the thickness of joints shall be between 6 to 15 mm. It shall be ensured while laying tiles that mud mortar rises vertically in joints to a height of about 15 mm. The joints shall be grouted after the brick tile are allowed to dry for a period of 24 hours. The joint shall be grouted with cement sand mortar (1 : 3). It shall be ensured that the joints are completely filled by mortar. The mortar shall be allowed to set for a minimum of 12 hours. Further pointing of the joints, to be done only if necessary. Before pointing, the grouted joints shall be brushed clean with a soft brush.

- **Curing :** The surface of finished roof shall be kept wet for a period of not less than 7 days.

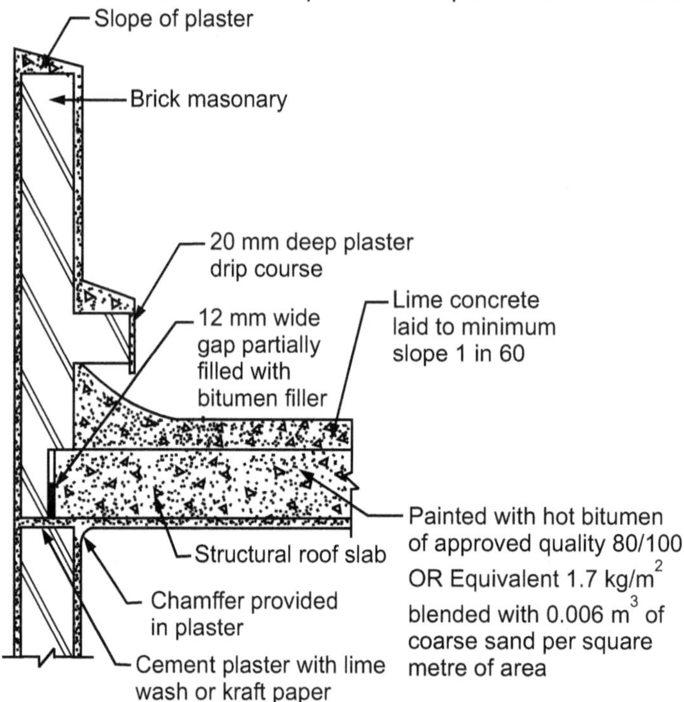

Fig. 5.5 : Lime concrete

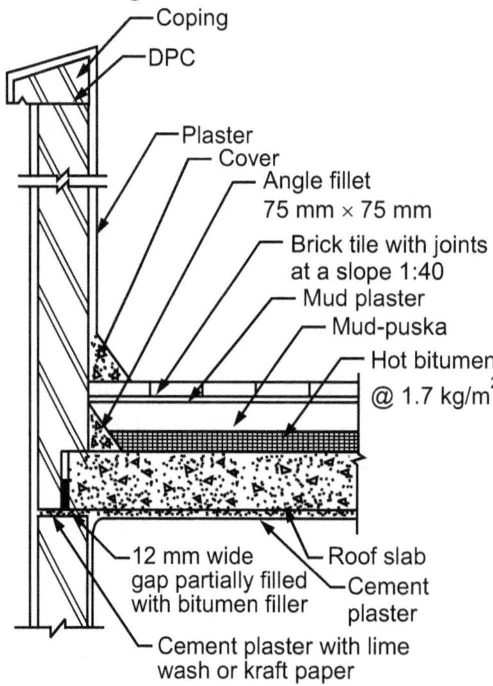

Fig. 5.6 : Mud phuska

5.2.4.2 Lime Concrete Terracing

- It comprises of laying, smashing and compacting of lime concrete and accomplishing the ideal incline with the goal that the water can be effectively depleted off. This framework can be utilized for both, new well as as old rooftops.

(A) Material :

- Lime concrete utilized in this framework is a blend of hydrated fat lime (according to Indian standard : 712),
- Pozzolanic material i.e. calcinated clay pozzolana (according to Indian standard : 1344), and
- Broken burnt clay bricks (as per IS : 3068) or natural stone aggregate (as per IS : 383) having greatest size 25 mm.

(B) Application Methodology :

- **Planning of lime concrete :** One part of slaked lime and two parts of burnt brick pozzolana, will be blended by volume. A watertight stage will be utilized for blending. Required amount of water will be sprinkled. The mix will be well ground in a plant or utilizing mechanical processor to acquire lime pozzolana mortar. Lime concrete is set up with coarse aggregate and lime pozzolana mortar, in extent of $2\frac{1}{2}$: 1 by volume. For hand blending, coarse aggregate will spread on to an even surface on the plank and lime pozzolana mortar will be at that point be equitably spread over it. Water will be added simply in adequate measure to the mix to acquire concrete of uniform consistency. The mixing will be done until all covered with mortar and a concrete of uniform consistency is achieved. If there should be an occurrence of machine mixing, first $2\frac{1}{2}$ parts by volume of clean saturated surface dried coarse aggregate will be added into the mixer and afterward the lime, by volume. Pozzolana mortar will at that point be added to the mixer and will be well mixed. Mixing will be proceeded until there is a uniform distribution of the material. Last alteration of water, to get concrete of required consistency might be made by including clear water.

- **Laying of lime concrete :** Prior to laying of lime solid, all preliminary work stated in IS 3067 will be finished. It comprise of cleaning of rooftop surface, filling craks with cement sand slurry. Rooftop surface ought to be made rough to create proper bond between lime concrete and base concrete. In the wake of cleaning, rooftop surface will be painted with hot bitumen of grade 80/100 @ 1.7 kg/m^2 mixed with coarse sand. Along these lines, laying of lime concrete will begin from an edge of a rooftop. It will further continue corner to corner towards focus and different sides considering the slope required for draining the downpour water easily. The normal thickness of lime concrete will not be under 100 mm. On flat roof, slope of 1 in 60 will be given. In case, of overwhelming precipitation territory, slant of 1 in 40 is suggested.

- **Ramming and compaction of lime concrete :** In the wake of lying it will be rammed with a rammer weighing not in excess of 2 kg and completed to the necessary uniformity and slope. Further consolidation will be done utilizing wooden 'thapies' with rounded edges. The labour will sit near one another and beat the surface gently. The beating will regularly must be carried on for seven days until the 'thapi' establishes no impression and bounce back promptly from it when struck. Ramming and compaction of lime concrete can be done mechanically so as to accomplish more progress. For this, a machine created by C.B.R.I., Roorkee might be utilized. During compaction, the surface will be sprinkled generously with lime mortar and little proportion of sugar solution. The sugar solution comprise of solution prepared by blending, 3 kg of jaggery, $1\frac{1}{2}$ kg of Bael Fruit to 100 liter of water by boiling) for getting improved water proofing of concrete. On culmination of beating, the mortar that goes ahead the top will be smoothened with a trowel or float.

- **Curing :** For lime concrete at least 10 days curing is required after compaction or until it solidifies. The curing is done by covering it with a thin layer of grass or straw, which will be kept wet persistently.

- **Treatment at intersection of rooftop and parapet** : Along the junction of rooftop surface with parapet wall, a strip of lime concrete fillet will be laid independently and completed smooth.
- **Finish :** On the off chance that the rooftop is open, one layer of burnt clay flat terracing tiles (as per IS : 2690 Part-1 and Part-2) might be laid over a thin layer of lime mortar. In any case, in the extraordinary condition for significant expansion and contraction, two layers of tiles might be put on the top of lime pozzolana concrete. These tiles ought to be joined with non- shrinking impenetrable concrete mortar.

5.2.4.3 Ferro Concrete Waterproofing

- This modern technique is completed in two layers of around 8 mm thickness for each layer. It comprises of wire mesh reinforced cement matrix layers which are vibrated *in situ*. The top surface additionally vibrated completed smooth with non-metallic fibres impregnation. The cement matrix and wire mesh layers are laid then again. Each cement matrix layer is laid on a bond coat for assured integrity. Each wire work layer is fixed with u-clips to the base.
- Since *in situ* low water cement ratio cement matrix is vibrated and become dense with immaterial pores. The wire mesh layers commonly two nos. forestall crack formation if at all because of shrinkage, temperature or structural relative movements. In this way the treatment is sturdy and perpetual. It takes ware and tare efficiently because of specific admixtures. The wire mesh layers add strength to the fundamental structure notwithstanding crack development counteractive action. It is established on insulated foundation layer of about normal thickness of 40 mm. There will be dead load decrease to conventional brickbat coba treatment. It adds solidarity to base structure.
- **Advantages of Ferro Concrete Waterproofing :**
    1. It cannot be punctured.
    2. It opposes mileage.
    3. It is changeless with nil/unimportant upkeep.
    4. It answer for ceaseless spillages.

5.2.4.4 Brick Bat Coba Waterproofing

- Laying cinder concrete 1 :15 (1 bond and 15 cinder of 13 mm and down gauge) in a normal thickness of 15 cm, the slope for the best possible drainage of rooftop being given to this layer.
- Laying 7.6 cm thick layer of lime concrete over the consolidated layer of cinder concrete, the lime concrete being set up by mixing 50% of mortar comprising of lime and sand mixed in the proportion of 1 : 2 with brick ballast 25 mm and down gauge.
- Spreading 13 mm thick layer of cement mortar 1 : 3 and laying tile bricks flat and open jointed over the mortar. At long last, grouting the joints in the bricks with cement mortar 1 : 3.

5.2.4.5 Waterproofing Utilizing Polymer Changed Bituminous Film

- Addition of polymer in bitumen improves its workability, penetration and softening behaviour, tensile and fatigue properties and ability to bridge movement of cracks/joints in the substrate. Non-woven fiberglass mat and non-woven polyester mat are commonly utilized as reinforcement to improve their attributes like lap joint strength, strength and flexibility. Polymer modified bituminous membrane are obtained by sandwitching non-woven polyester fabrics or fiberglass mat between layers of top notch polymer modified bituminous. These films have high softening point, high tensile strength, high tear and puncture resistance, high joint strength and low water vapour transmission. This framework is appropriate for new just as old rooftops.

Application Method :

1. **Provision of slope and cleaning of roof surface :** Before use of water proofing treatment rooftop surface ought to be furnished with a base slant of 1 out of 100 with plain cement concrete. After arrangement of slope every single preliminary stir like filling up of cracks by cement sand slurry, arrangement of sufficient number of drain outlets, arrangement of 75 mm filet at intersection of rooftop with parapet wall, provision of a groove/ chase in parapet wall and so forth as per IS :3067 ought to be finished.

2. Laying a layer of cold applied bitumen primer @ 0.2 to 0.4 l/sq. m. on whole rooftop surface.

3. Laying 85/25 grade hot blown bitumen @ 1.2 kg/sq.m. everywhere throughout the surface.

4. Laying 2.5-3 mm thick polymer modified bituminous layer with non-woven polyester fiberglass mat reinforcement, applied by torch with fixing every one of the joints.

5. Laying 85/25 evaluation hot blown bitumen @ 1.2 kg/sq. m. everywhere throughout the surface.

6. Last finish with china mosaic tiles on a 15 mm thick grey cement plaster bed.

7. Technique for application may somewhat vary contingent on item and makers' proposals. As there is no pertinent IS code of practice for this water proofing system, work ought to be completed according to producers' suggestions. Clients are encouraged to gather total literature from makers and concentrate totally before use of treatment.

5.2.4.6 Application of Polymer Modified Cementitious Slurry Coating

- Following steps are involved in application of this system.

  1. **Planning of rooftop surface :** The surface will be cleaned to expel all residue, remote issues, lose materials or some other stores of tainting. Breaks and despondency will be topped off by fillers. Arranged surface will be altogether pre wetted for 60 minutes.

  2. **Planning of polymer altered cementitious slurry :** Dry mix and fluid mix cannot avoid being blend into the ideal proportion according to proposal of provider. The blend will be mixed completely, until no air pockets stay in the blend. Any bump found in blend will be expelled.

  3. First layer of polymer changed cementitious slurry will be applied by brush on wet cleaned surface.

  4. Along these lines, fiberglass material will be laid over first layer of polymer altered cementitious slurry.

  5. Second layer of polymer altered cementitious slurry will be laid over fiber glass material.

  6. Polymer altered cementitious brush beating will be applied over second layer of polymer adjusted covering.

  7. On brush topping, screed concrete, 1 :2 :4 admixed with suitable integral water proofing compound 25 to 40 mm thick to a min. slope of 1 in 100 with aggregate size down 10 mm with maximum water cement ratio 0.45, shall be laid.

- Above system may slightly differ from case to case depending upon the instruction of supplier of water proofing system. There is no relevant Indian standard/ other code of practice for this system. Therefore, work should be carried out as per manufacturers'/ suppliers instructions. Users are advised to collect complete literature from manufacturer and study carefully prior to application of treatment. Since, there are no relevant Indian or any other standards available, this system needs to be adopted carefully.

5.2.4.7 Treatment of Sunk Floor in Toilets/Kitchen

Type I :

Step-1 : Remove all materials /flooring from the sunken floors and expose the drainage pipes/G.I. water supply lines.

Step-2 : Test the G.I. water supply lines less than 6 kg per sq. cm. Water pressure using the pressure-testing machine with pressure guage, which is promptly accessible in the market.

Step-3 : Test the drainage pipes and other joints for leakages if any, by plugging of horizontal pipe at tee junction with vertical stack and filling with water upto the finished floor level for 48 hours. Leakages assuming any, will be taken care of.

Step-4 : Provide 40 mm diameter G.I. pipe spout and CC flooring with water proofing compound laid in slope (1 : 48 minimum) for draining out leaking water, if any, from the sunken portion in the shaft.

Step-5 : Provide 12 mm thick cement plaster 1 : 3 (1 cement : 3 fine sand) mixed with water proofing compound on the vertical walls of the sunken portion including providing necessary repair around the drainage spout provided.

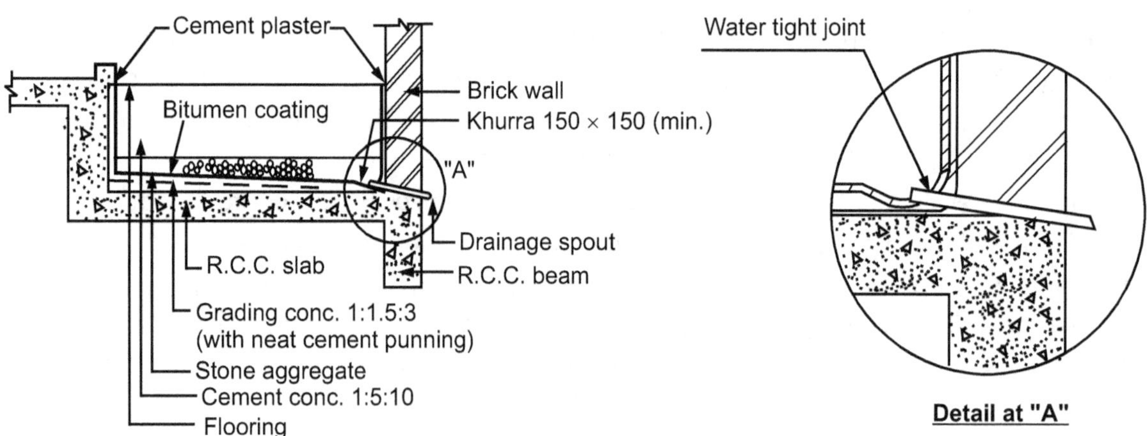

Fig. 5.7 : Treatment of sunk floor in toilets/kitchen

Type II :

Step 1 : Provide drainage spout.

Step 2 : Lay concrete 1 : 1.5 : 3 to slant towards draining spout with floating coat of neat cement.

Step 3 : Provide and lay water supply and draining pipes and test for leakage. The drainage /sanitary pipes shall be laid to slope and a clear gap underneath the pipes shall be made available by providing suitable CC blocks at appropriate location for free flow (of leaked water, if any) towards drainage spout.

Step 4 : 18 mm thick cement plaster 1 : 3 on vertical faces mixed with polymer modified with water proofing compound and floating coat of polymer modified neat cement slurry and coated with bitumen (2 coats).

Step 5 : 150 mm thick 20 mm single sized stone aggregate to provide a filter media for any leaking water to travel to drain spout below c.c. 1 : 5 : 10 (1-cement : 5-coarse sand : 10-40 mm and downsized stone aggregate).

Step-6 : Provide two coats of bitumen (85/25 grade) coating in the sunken portion to guarantee that the whole surface is appropriately covered and drainage pipes are also painted with bitumen.

Step-7 : Provide a 100 mm thick dry stone aggregate in the sunken portion. The mouth of the drainage spout provided earlier shall have graded filter so as not to get choked.

Step-8 : Rest of the depth of sunken portion shall be filled with lean concrete 1 : 5 : 10 (1 cement : 5 coarse sand : 10 aggregate) with stone aggregate and flooring laid to slope.

Step-9 : The rusted/leaking G.I. pipes wherever noticed during carrying out the repairs shall be replaced or otherwise be cleaned of rust and coated with polymer modified cement slurry and two coats of bitumen painting.

- Screed concrete of 1 : 2 : 4 mix with reasonable essential water proofing compound will be laid. It shall be 25 to 40 mm thick to a minimum slope of 1 in 100 with total size down 10 mm with W/C ratio proportion 0.45. Above system may somewhat contrast from case to case contingent on the guidance of provider of water proofing system. There is no applicable Indian standard/other code of practice for this system. Hence, work ought to be done according to makers'/providers guidelines. Clients are encouraged to gather total writing from maker and concentrate cautiously before utilization of treatment. Since, there are no important Indian or some other principles accessible, it should be embraced cautiously.

5.3 REPAIR OPTIONS

5.3.1 Epoxy Injection

- The injection of a low viscosity epoxy is a possible repair method for cracks. The epoxy injection to be effective, the crack must be free of dirt grease or other contaminations.

- The crack shall be between 0.02 mm and 6 mm in width. For very fine crack, less than 0.15 mm in width, the entry ports should be spaced not more than 150 mm apart.

- It is necessary to choose carefully to match the individual job requirement.

- The capability of bonding to moist concrete, shrinkage thermal and elastic properties of hardened resins and other special needs such a fire resistance high temperature stability.

- It is relatively new work, satisfactory cleaning can often be achieved by vacuum cleaning, a head of the sealing operation. Compressed air or blasting with water or air/water mix have been suggested but the process tends to drive dust and contaminations into the bottom of cracks.

- Injection should start at the lowest post and be continued until resin appears at the next higher port. The injection nozzle is then removed from the port seal and the nozzle moved upto the next port. If the pumping pressure cannot be maintained in a cracks that appears full, the epoxy resin may leak.

- Finally when the injected resin is cured, the sealing adhesive must be removed by grinding, cutting and at port must be made good with epoxy. The cut width of about 20 mm to be filled with dry pack or epoxy mortar.

Three methods of providing entry ports :

- Drilled holes with fitting inserted and bonded in, with the adhesive used for sealing.

- Bonded flush fitting attached by means of the sealing adhesive.

- Interrupted seal using a gasket that covers the unsealed portion.

5.3.2 Patch Repairs

- **Mortar for Repair Cracks :**

1. **Symptoms of defects :**

    (a) Surface and body of concrete.

    (b) Spalling, rust and dampness stain on the surface.

    (c) Corrosion of reinforcement, porous concrete near surface.

    (d) Non-conformation of surface, shape and size of member.

2. **Repair Materials :**

    (a) Patching material.

    (b) Plain cement mortar epoxy resin mortar.

    (c) Polymer modified mortar, polyester resin mortar.

3. **Crack Repair Methods :**

    (a) Epoxy injection grooving and sealing, grouting.

    (b) Flexible sealing, poly-impregnation, dry packing.

    (c) Overlays and surface treatment.

- This process of natural crack repair is called auto generous sealing. Healing depends upon the calcium hydroxide in cement paste, carbon-di-oxide present in the atmosphere and presence of moisture.

- **Steps involved :**

    1. Preparation of crack, drilling holes, clearing and drying of cracks.

    2. Sealing of cracks surface, fixing of injection port in holes.

    3. Mixing of epoxy resin, injection of mixed epoxy resin in port.

    4. Removal of ports and plugging the holes, removal of surface sealing and finishing the surface.

Procedure for Repair Cracks in Concrete with a Concrete Patch :

1. Chisel out the crack to create a backward-angled cut, using a cold chisel and a hammer.
2. Clean loose material from the crack using a wire brush, or a portable drill with a wire wheel attachment.
3. Apply a thin layer of bonding adhesive to the whole repair area using a paint brush. The bonding adhesive helps to keep the repair material from loosening or popping out of the crack.
4. Mix vinyl reinforced patching compound and trowel it into the crack. "Feather" the repair with a trowel, so it is even with the surrounding surface.

Large Crack Variation :

- Chisel out the crack, and clean loose material from the crack. Then pour sand into the crack within the surface. Prepare sand-mix concrete, adding a concrete fortifier, and trowel the mixture into the crack. Feather until leveled with the surrounding surface using a trowel.

5.3.3 Mortar and Dry Pack

- Dry pack is suitable for filling holes whose depth is at least equal to the smallest surface dimension of the repair area. The holes should be at least 25 mm deep. Dry pack is not suitable for shallow depressions. The holes that go right through concrete section where the filling cannot be properly rammed.

- Dry pack mortar is normally a mix of one part of OPC to 2.5 part of fine sand. The correct measure of water will produce a mortar which is at the point of becoming rubbery when it is solidly packed. Less water will not make a sound pack as it cannot be properly rammed. More water lead to shrinkage and a loose repair.

- The holes should be prepared so that they are sharp and square at the surface edge. The interior surface ought to be roughed and if possible under cut marginally. All repairs and defective concrete must be removed and the surface of the hole left clean. Dry should be packed in layer which has a compacted thickness of about 10 mm. The compacting efforts should be directed at a slight angle towards the sides of the hole. The holes should not be overfilled and can be done by pounding on a piece of hardwood laid on the surface.

- Holes are vertical or overhead surface or not likely to be repaired effectively by this method and epoxy mortar is may be needed. The region to be repaired should be cleaned and roughened and kept wet for few hours. The repair mortar should be mixed to a plastic consistency. A small quantity of cement mortar should be scrubbed into the surface with a wire brush. The repair mortar compacted thoroughly should be tight filling around the edges of holes. Curing should be applied as soon as possible and kept in place for at least 7 days. There should be good bond between old concrete and repair. The expansive cement or admixtures have been advocated for replacement mortar repair. Any admixture used should not be relies on the corrosion of iron fillings. Expansive admixtures are used to grouting purpose.

- Epoxy method is used for repair jobs however its properties and performance are very much dependent on the skill of crews, types of equipment used and the conditions under which placing is carried out. It has more cost, availability of equipment on operation features. Possible admixture is an effective material for replacing defective concrete. The durability is highly dependent on the preparation of the bonding surface and skill of nozzle man. Air entraining admixture with dry mix shotcrete in the hope that will provide additional insurance of durability. Two major modifications shotcrete have been introduced in the addition of fibres and more recently the addition of silica fume.

- The addition of silica fume to shotcrete has produce extraordinary benefits in the properties of the plastic and the hardened materials. Dry mix silica fume shotcrete commonly has a 28 days compressive strength of up to 60 Mpa. The addition of silica fume to fibre shotcrete produce more flexural strength.

Table 5.2 : Concrete repair : 'Methods' and 'Materials'

Defects	Repair Methods	Materials

• Live Cracks	– Caulking
	– Pressure injection with 'flexible ' filler
	– Jacketing :
	   * Strapping
	   * Overlaying
	– Strengthening
		Elastrometric sealer.
		'Flexible' epoxy (resin and hardener mix) filler.
		
		Steel wire or rod
		Membrane or special motar
		Steel plate, post tensioning, stitching etc.

• Dormant Cracks	– Caulking – Pressure injection with 'rigid' filler – Coating – Overlying – Grinding and Overlay – Dry-pack – Shotcrete/Gunite – Patching – Jacketing – Strengthening – Reconstruction	Cement grout or mortar, Fast-setting mortar. 'Rigid' epoxy (resin and hardener mix) filler Bituminous coating, tar Asphalt overlay with membrane Latex modified concrete, highly denser concrete Dry-pack Mortar (cement), Fast-setting motar Cement mortar, Epoxy or Polymer concrete Steel rod Post tensioning, etc. As needed
• Voids • Hollows • Honeycombs	– Dry pack – Patching – Resurfacing – Shotcrete/Gunite – Preplaced aggregate – Replacement	Dry-pack Portland cement grout, mortar, cement Epoxy or Polymer concrete Fast-setting mortar Coarse aggregate and grout As needed
• Scaling Damage	– Overlaying – Grinding – Shotcrete/Gunite – Coating – Replacement	Portland cement concrete, Latex modified concrete Asphalt cement, Epoxy or polymer concrete Fast-setting mortar, Cement mortar Bituminous, Linseed oil coat, Silane treatment As needed
• Spalling Damage	– Patching – Shortcrete/Gunite – Overlay – Coating – Replacement	Concrete, Epoxy, Polymer, Latex, Asphalt Cement mortar, Fast-setting mortar Latex modified concrete, Asphalt concrete, Concrete Bituminous, Linseed oil, Silane etc. As needed

5.3.4 Grooving and Sealing

- This is the simplest and most common method of crack repair. It can be executed with relatively unskilled personnel and can be used to seal both fine pattern cracks and larger isolated cracks.

- The system can be used to repair dormant cracks that are of no structural significance, and is used to seal the cracks against the ingress of moisture, chemicals and carbon dioxide. This involves enlarging the crack along its exposed face and sealing it with crack fillers as shown in Fig. 5.8. Care should be taken to ensure that the entire crack is routed and sealed.

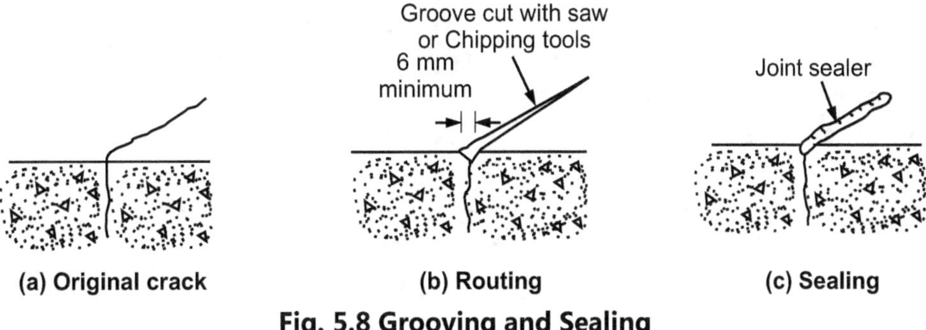

Fig. 5.8 Grooving and Sealing

5.3.5 Stitching

- In this technique, the crack is bridged with U-shaped metal units stitching dogs before being repaired with a rigid resin material. This can establish restoration of the strength and integrity of cracked section; due care is to be given to make analysis check to ensure that this will perform well under applied loads as shown Fig. 5.9. A non-shrink or an epoxy resin based adhesive should be used to anchor the legs of the dogs.

- Stitching is suitable when tensile strength must be reestablished across major cracks, although stitching will not close the crack, and it is way of stopping the movement of active crack and thereby preventing it from spreading. Stitching dogs should be of variable length and orientation and so located that the tension transmitted across the crack is not applied to a single plane within the section but us spread over an area.

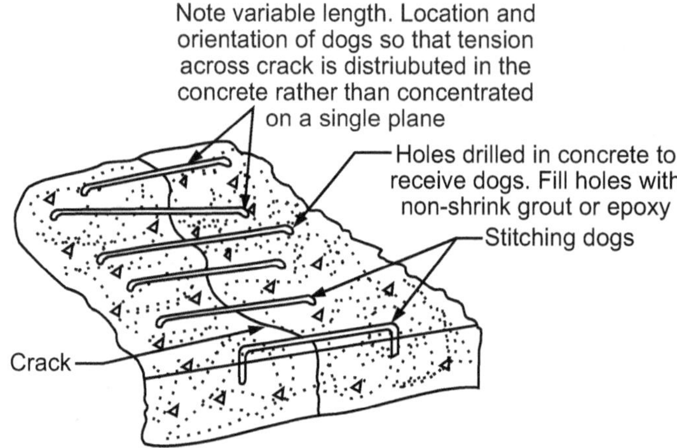

Fig. 5.9 : Stitching

5.3.6 Bonding

- Cracks in concrete may be bonded by the injection of epoxy bonding compounds under pressure. A usual practice is to drill into cracks from face of the concrete at several locations. Water or a solvent is injected to flush out the defect. The surface is then allowed to dry. The epoxy is injected into the drilled holes until it flows out through the other holes.

- The epoxy is injected into the drilled holes until it flows out through the other holes. Bonding with epoxies-cracks as narrow as 0.075 mm can be sealed with epoxy compounds, usually pressure injection is restored to seal the cracks.

5.3.7 Bandaging

- A flexible strip is fixed over the crack with only the edged of the strip bonded. Where movement is not all in one plane, where excessive movement beyond that which can be accommodated by a recess of convenient size, or if there are factors which prohibit the cutting of a recess, a surface bandage can be used. In areas which are subject to traffic, the flexible bondage will be coated over with a wearing course.

5.3.8 External Stressing

- Cracks can be closed by inducing a compressive force, sufficient to overcome the tension and to provide a residual compression. The principle is very similar to stitching, except that the stitches are tensioned; rather than plain bar dogs which apply no closing force to the crack. Some form of abutment is needed for providing an anchorage for the prestressing wires or rods.

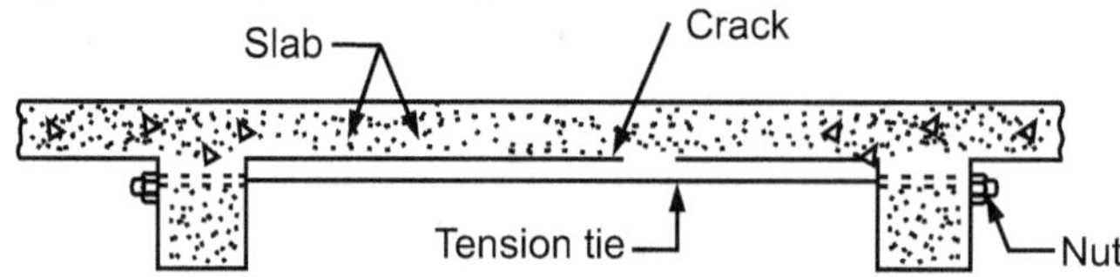

(a) To correct cracking in beams

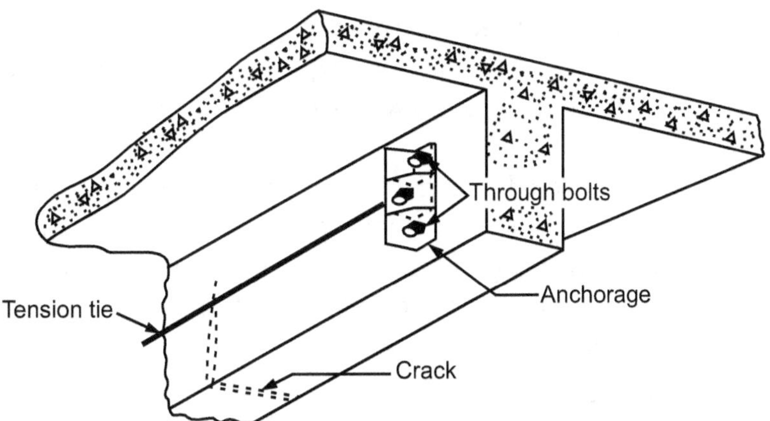

(b) To correct cracking in slabs

Fig. 5.10 : External stressing

5.3.9 Grouting

- Grouting is defined as, "injection of liquid grouting material into the concrete, masonry, soil etc. environment under pressure. During the grouting process, cracks and pores are filled with the grouting material, which subsequently hardens. Grouting can be performed in a similar manner as the injection of an epoxy. However the use of an epoxy is the better solution except where considerations for the resistance of cold weather prevent such use in which case grouting is the comparable alternative.

- **Precautions while Grouting for Repair of Cracks in Concrete Structures :**

  1. Holes are drilled in structure along cracks and in an around hollow spots.

  2. If there are several cracks, holes can be drilled in as staggered manner at 500 to 750 mm spacing in both directions covering adequately the area proposed to be grouted. Holes spacing can be altered as per site conditions.

  3. G.I. pieces (12 to 20 mm dia. × 200 mm) with one end threaded or PVC nozzles are fixed in the holes with rich cement mortar.

  4. All the cracks and annular space around G.I. pipes are sealed with rich cement mortar.

  5. All the cracks are cut open to a 'V' shaped groove, cleaned and sealed with rich cement mortar.

  6. All the grout holes should be sluiced with water using the same equipment a day before grouting as per following sequence; so as to saturate the masonry.

- All holes are first plugged with proper wooden plugs or locked in the case of PVC nozzles. The bottom most plug and the two adjacent plugs are removed and cement-water/chemical injected in the bottom most hole under pressure.

- When the cement-water/chemical comes out through the adjacent holes the injection of cement-water/chemical is stopped and the plugs in the bottom most hole and the one immediately above are restored. The process of grouting of concrete cracks is repeated with other holes till all the holes are covered. On the day of grouting, all the plugs are removed to drain out excess cement-water/chemical and restored before commencing grouting. The grout is kept fully stirred/ agitated under pressure throughout the grouting.

- The grouting is carried out till refusal and/or till grout starts flowing from the adjacent hole. A proper record of the quantity of grout injected into every hole should be maintained.

- After grouting, curing should be done for 14 days. Only such quantities of material for preparing grout should be used, as can be used within 15 minutes of its mixing. Grouting equipment must be cleaned thoroughly after use.

- **Precautions during Grouting of Cracks in Concrete :**
  1. Immediately after grouting work, all the grouting equipment including the slurry and mixing drums, pipes, nozzles, etc. should be thoroughly washed so that set cement does not damage the equipment.
  2. After the work has been completed, it should be inspected thoroughly and should be kept under observation for a period of 6 months to 12 months for its behaviour after grouting.
  3. In case arch masonry of bridges is grouted to strengthen the structure, some load tests may be carried out in selected cases to satisfy that grouting has helped to reduce the deflection of crown and spread at the springing to within permissible limits.

5.3.10 Guniting/Shotcreting

- **Gunite :** Gunite can be defined as, "mortar conveyed through a hose pneumatically protect at a high velocity on to a surface". The method has been further developed by introducing small size coarse aggregate into mix. This process is made economical by reducing the cement content. The force of jet impacting on the surface compact the material. Use of accelerator to assist over knead hacking is practiced.

- **Shotcreting :** The newly developed ready set cement can also be used for shortening process. The process is mostly used for application of mortar of less thickness. Shotcreting is similar principle of guniting for achieving greater thickness with small coarse aggregate.

- **Dry mix process :** Cement and sand are thoroughly mixed. The cement and sand mixer is feed into special air pressurised mechanical feeded as 'gun'. The mixer is melted into delivery hose by a distributor within the gun. This material is carried by compressed air through the delivery hose to a special nozzle. The water is spayed under pressure and intimately mixed with sand cement jet. The wet is jetted from the nozzle at high velocity on to the surface to be gunited. The dry-mix methods make use of high velocity or low velocity system. The high velocity gunite is produced by using a small nozzle.

- **Wet mix process :** To produce a high nozzle velocity about 90-120 m/s. This result is in exceptional good compaction. The lower velocity gunite is produced using large diameter hose for large output. The compaction will not be very high. In the wet-mix process the concrete is mixed with water as for ordinary concrete before conveying through the delivery pipe line to the nozzle, at which point it is jetted by compressed air, onto the work in the same way, as that of dry mix process. The wet-mix process has been generally discarded in favors of the dry-mix-process, owing to the greater success of the latter.

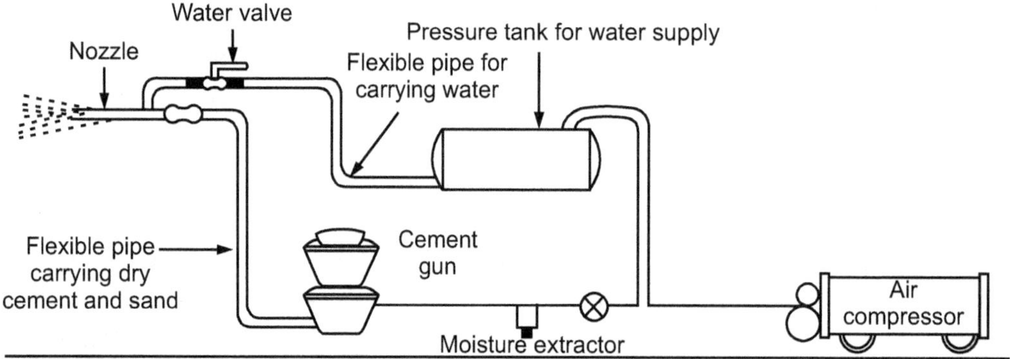

Fig. 5.11 : General arrangement of apparatus in gunite system

- **Use of Shotcrete :**
  1. The high cost to shotcrete limits is application to certain special circumstance. The saving in shuttering cost makes it particularly applicable for this section. It will bond extremely well to the existing concrete to masonry and to exposed rock.
  2. Suitably prepared steel surface also can be covered with gunited concrete.

3. It is difficult to removes rebound materials as it tends.

4. Defects of this type can result in porous concrete and also contribute to high permeability.

5. It is difficult to obtain a satisfactory surface finish with shotcrete.

6. The application of the shotcrete process is limited to exception areas.

7. Admixtures can be used in shotcrete to produce the same effects as in ordinary concrete.

8. The drying shrinkage will depend on the water content. Shrinkage and creep of wet shotcrete is likely to be high.

9. The durability or resistance to frost action and other agencies of dry shotcrete is good.

10. One of the strong point of shotcrete is its excellent bond with old concrete, rock (face) phase, metal sheet.

11. The use of shotcrete is frequently adopted for tunnelling operation.

12. Generally it should have quick setting properties. This properties are usually obtained by the use of powerful accelerator in the mix. This accelerator should be compactable with the cement and concrete with respect to durability and stability.

13. Use of fibre reinforced shotcrete is one of the recent innovation. Fibre reinforced shotcrete process increase the tensile strength of the shotcrete.

14. Another important innovation made is the polymer shotcrete aggregate and monomer are mixed together.

5.3.11 Fibre Wrap Technique

- The fibre wrap technique, also known as Composite Fiber System is a non-intrusive structural strengthening technique that increases the load carrying capacity (shear, flexural, compressive) and ductility of reinforced concrete members without causing any destruction or distress to the existing concrete.

- There are two systems followed in adopting this technique :

    1. **Bi-directional Woven Fabric :** This system comprises of woven fabric presoaked in specially formulated epoxy and applied over prepared surface after application of epoxy primer. Woven fibre fabric is composed of bi-directional high strength fibers that are combined with specially formulated epoxy in a pre-determined proportion to form a composite-material. This composite material is wrap applied onto the reinforced concrete or steel member requiring strengthening or protection and left to cure at ambient temperature. The subsequent layer/s of unidirectional fibre fabric could be applied after giving the required overlap along the direction of fibres as per design requirements.

    2. **Uni-directional E-glass Fibres :** This system comprises of precut unidirectional E-glass fibre wrapped over epoxy primer applied prepared surface of member requiring structural strengthening and/or surface protection. Subsequent to its wrapping, it is saturated with epoxy using rollers and stamping brushes manually to remove air bubbles, if any and left to cure at ambient temperature. The subsequent layer/s of unidirectional fibre fabric could be applied after giving the required overlap along the direction of fibres as per design requirements.

- Though the underlying principle of the above two methods is more or less identical, but the application techniques and basic materials adopted are at slight variance. Each of the above systems has its own merits.

- Enhancement in lateral drift ductility and horizontal shear carrying capacities of a concrete member can also be obtained by confinement of the member by this method. The flexural, shear and axial load carrying capacities of the structural members can be enhanced by appropriate orientation of primary fibres of the composites. The resulting cured membrane not only strengthens the reinforced concrete member but also acts as an excellent barrier to corrosive agents, which are detrimental to concrete and the reinforcement. Ingress of water, oxygen and carbon dioxide through the external surface of concrete member is prevented by the application of composite jacket.

- The system is useful for its structural enhancement and protection capabilities under severe environmental conditions. It can be used for retrofitting of a wide variety of structures that include bridges, flyovers, chimneys, water tanks, buildings, large diameter pipes, industrial plants, jetties, sea-front and underwater structures.

5.4 REPAIR OF CORRODED RCC ELEMENTS

- **Corrosion** is defined as, "the process of deterioration (or destruction) and consequent loss of a solid metallic material, through an unwanted (or unintentional) chemical or electro-chemical attack by its environment, starting at its surface". Thus, corrosion is a process of "reverse of extraction of metals".

- Corrosion of steel concrete is an electro-chemical process. When there is a difference in electrical potential, along the reinforcement in concrete, an electro-chemical cell is set up. In the steel, one part becomes anode (an electrode with a +ve charge) and other part becomes cathode, (an electrode with a negative charge) connected by electrolyte in the form of pore water, in the hardened cement paste. The positively charged ferrous ions (Fe^+) at the anode pass into solution, while the negatively charged free electrons pass through the steel into cathode, where they are absorbed by the constituents of the electrolyte, and combine with water and oxygen to form hydroxyl ions (OH). These travel through the electrolyte and combine with the ferrous ions to form ferric hydroxide, which is converted by further oxidation to rust.

5.4.1 Corrosion Inhibitors

- A corrosion inhibitor is an admixture that is used in concrete to prevent the metal, embedded in concrete form cording. There exists various types of inhibitors like cathode, anode, mixed and dangerous, safe.

- Of the available corrosion inhibiting materials, the most wide used admixture is based on calcium nitric. It is added to the concrete during mixing of concrete. The typical dosage is of the order of 10-30 litres per m^3 of concrete, depending on the chloride levels in concrete. In the high pH of concrete the steel is protected by a passivating layer of ferric oxide, on the surface of steel. Passivation may be defined as, "phenomenon in which a metal or an alloy exhibits a much higher corrosion resistance, than expected from its position in the electrochemical series".

- Passivity is the result of the formation of a highly protective, but very thin, (about 0.0004 mm thick) and quite invisible film on the surface of metal or an alloy, which makes it more fine). However, the passivation layer also contains some ferrous oxide, which can initiate corrosion, when the chloride ions reach the steel. The nitrite ions present in the corrosion inhibiting admixture will oxidize the ferrous oxide passivation layer even in the presence of chlorides. The concentration of nitrite must be sufficient, to cope up with the continuing ingress (entrance) of chloride ion. Calcium nitrite corrosion inhibitor comes in a liquid form, containing about 30% calcium nitrite solids by weight. The more corrosion inhibitor is added, the longer the onset of corrosion will be delayed.

- Since most structures in a chloride environment reach a level of about 7 kg of chloride iron per m^3 during their service life, use of less than 18 litres/m^3 of calcium nitrite solution is not recommended.

- Without an inhibitor, the reinforcing steel starts to corrode, when the chloride content at the rebar reaches a threshold level of 0.7 kg/m^3. Although the corrosion process starts when the threshold level is reacted, it may take several years for staining, cracking and spalling to become apparent, (clear) and several more years before deterioration occurs. Adding calcium nitrite increases this corrosion threshold. When you ass 20 litres/m^3, corrosion will not begin until over 7.7 kg/m^3 of chloride is present in the concrete at the rebar.

5.4.2 Cathode Protection

- Cathode protection is one of the effective, well known, and extensively used methods for prevention of corrosion in concrete structures in more advanced countries. Due to high cost and long term monitoring required for this method, it is not very much used in India.

- The cathode protection comprises of application of impressed current, to an electrode laid on the concrete, above steel reinforcement. This electrode serves as anode and the steel reinforcement, which is connected to the negative terminal of a DC source acts as a cathode. In this process, the external anode is subjected to corrode and the cathode reinforcement is protected against corrosion and hence the name 'Cathode Protection". In this process, the $-ve$ chloride ions, which are responsible for the damage of the passivating film, are drawn away from the vicinity of steel towards the anode, where they are oxidized to form chlorine gas. The other recent development in corrosion control methods are Re-alkalization and Desalnation.

- The re-alkalisation process allows to make the concrete alkaline again and passivate the reinforcement steel, by electro-chemical method. This brings back the lost alkalinity of concrete of sufficiently high level to reform and maintain the passive layer on the steel. In the desalination process, the chloride ions are removed from the concrete, particularly from the vicinity of the steel reinforcement by certain electrical method to re-establish the passive layer of the steel.

- It appears that, the application of cathodic systems for protection of concrete structures, offers some real hope to the concrete technologist, but the field remains open for the introduction of innovative methods to overcome problems of both technique and cost.

5.4.3 Corrosion Resistant Steel

- It is found that susceptibility of mild steel to corrosion is not significantly affected by composition, grade or level of stress. Hence, substitute steel for corrosion resistance must have a significantly different composition. Based on some success in atmospheric corrosion, weathering steels of the corten type were tested in concrete. They did not perform will in moist concrete, containing chlorides. It is observed that weathering corrode in similar concrete environments, to those causing corrosion of high-yield steel. They noted that although the total amount of corrosion was less, than would occur on high-yield steel under similar conditions, deep localized pitting developed, which could be more structurally weakening.

- Stainless steel reinforcement has been used in special applications, especially as fitments in precast members, but is generally too expensive to use as a substitute for mild steel. Very high corrosion resistance was shown by austenitic stainless steel in all the environments, in which they were tested, but the observations of some very minor printing in the presence of chlorides lead to the warning that crevice corrosion susceptibility was not evaluated in the test program. High titanium alloy bar is being used in some countries. This bar is grouted into holes, drilled into the marble stabs, and the grouts are based either on Portland cement or Epoxy.

5.4.4 Coatings for Steel

- The object of coating to steel bar is to provide a durable barrier to aggressive materials, such as chlorides. The coatings should be robust to withstand fabrication of ribcage, and pouring of concrete and compaction by vibrating needle. Simple cement slurry coating is a cheap method for temporary protection, against rusting of reinforcement in storage.

- Central Electro Chemical Research Institute (CECRI), Karaikudi, have suggested a method for prevention of corrosion in steel reinforcement in concrete. The steps involved in this process are :

  1. **De-rusting :** The reinforcement are cleaned with a de-rusting solution. This is followed without delay by leaning the rods, with wet waste cloth and cleaning powder. The rods are then rinsed in running water and air dried.

  2. **Phosphating :** Phosphate jelly is applied to the bars, with fine brush. The jelly is left for 45-60 minutes, and then removed by wet cloth an inhibitor solution is then brushed over the phosphated surface.

  3. **Cement Coating :** Slurry is made by mixing the inhibitor solution with portland cement and applied on the bar. A sealing solution is brushed after the rods are air cured. The sealing solution has an insite curing effect. The second coat of slurry is then applied and the bars are air dried.

4. **Sealing :** Two coats of sealing solution are applied to the bars, in order to seal the micro-pors of the cement coated and to make it impermeable to corrosive soils. This is patent method evolved by CECRI, and licence is given to certain agencies. It is one of the effective methods of coating rebars. The fusion bonded epoxy coating is a specialized job, carried out in a factory, and not at site of work. Plants are designed to coat the straight bars in a continuous process.

5. Although epoxy coated bars have an excellent protection to corrosion in aggressive environment, there are a few limitations. After the treatment, cutting and bending may injure the steel, which needs certain site treatment. The site treatment is likely to be inefficient. The presence of any defect in the treated body can induce severe localized corrosion, which defeated very purpose. The bars cannot be welded. The epoxy is not resistant to rays of sun. The bars should not be exposed to sun for long duration before use. The coating may get damage during vibration of concrete. The treatment is very costly, as that of steel. This method of protection to the steel is being given to all the flyovers, and other structures at Mumbai.

6. **Galvanized Reinforcement :** Galvanizing of reinforcement consists of dipping the steel bars in molten zinc. This results in a coating of zinc, bonded to the surface of steel. The zinc surface with calcium hydroxide in the concrete, to form a passive layer and prevents corrosion.

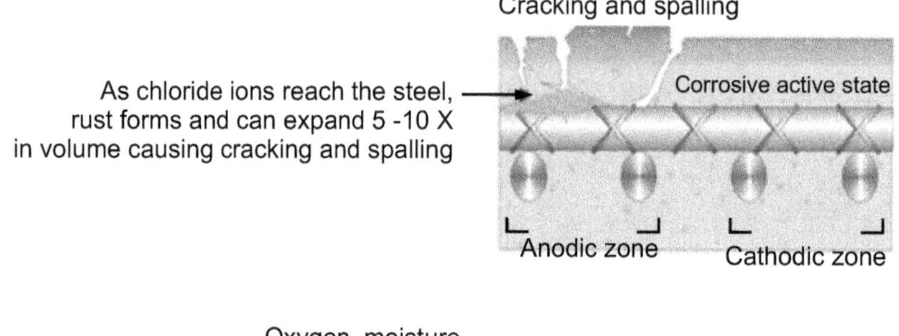

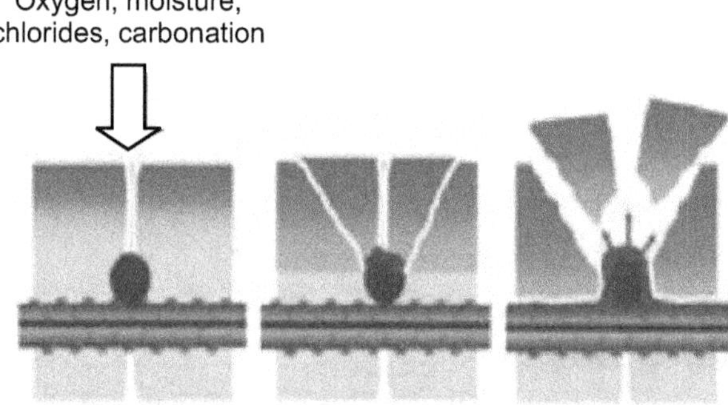

Fig. 5.12 : Cracking of concrete due to corrosion

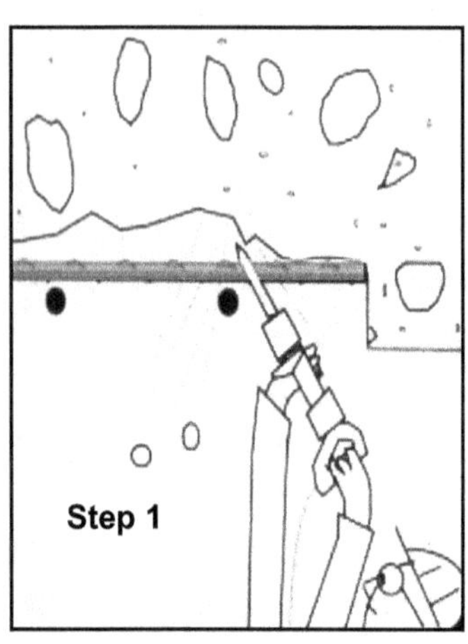

- By using a marker pen, delimit the area of damage.
- Break beyond the delimited area until reaching sound concrete and steel.
- With a jackhammer, saw-cut 2 cm deep in 90° angle the perimeter of breaking, in order to avoid feather edges.
- All weak, damaged and easily removable concrete should be chipped away. If the re-bars are only partially exposed after all unsound concrete is removed, it may not be necessary to remove additional concrete to expose the full circumference of the reinforcement. When the exposed reinforcement steel has a loose wrap, corrosion is not well bonded to the surrounding concrete. The concrete removal should continue until a clear space of 15 to 25 mm is created behind the reinforcing steel.

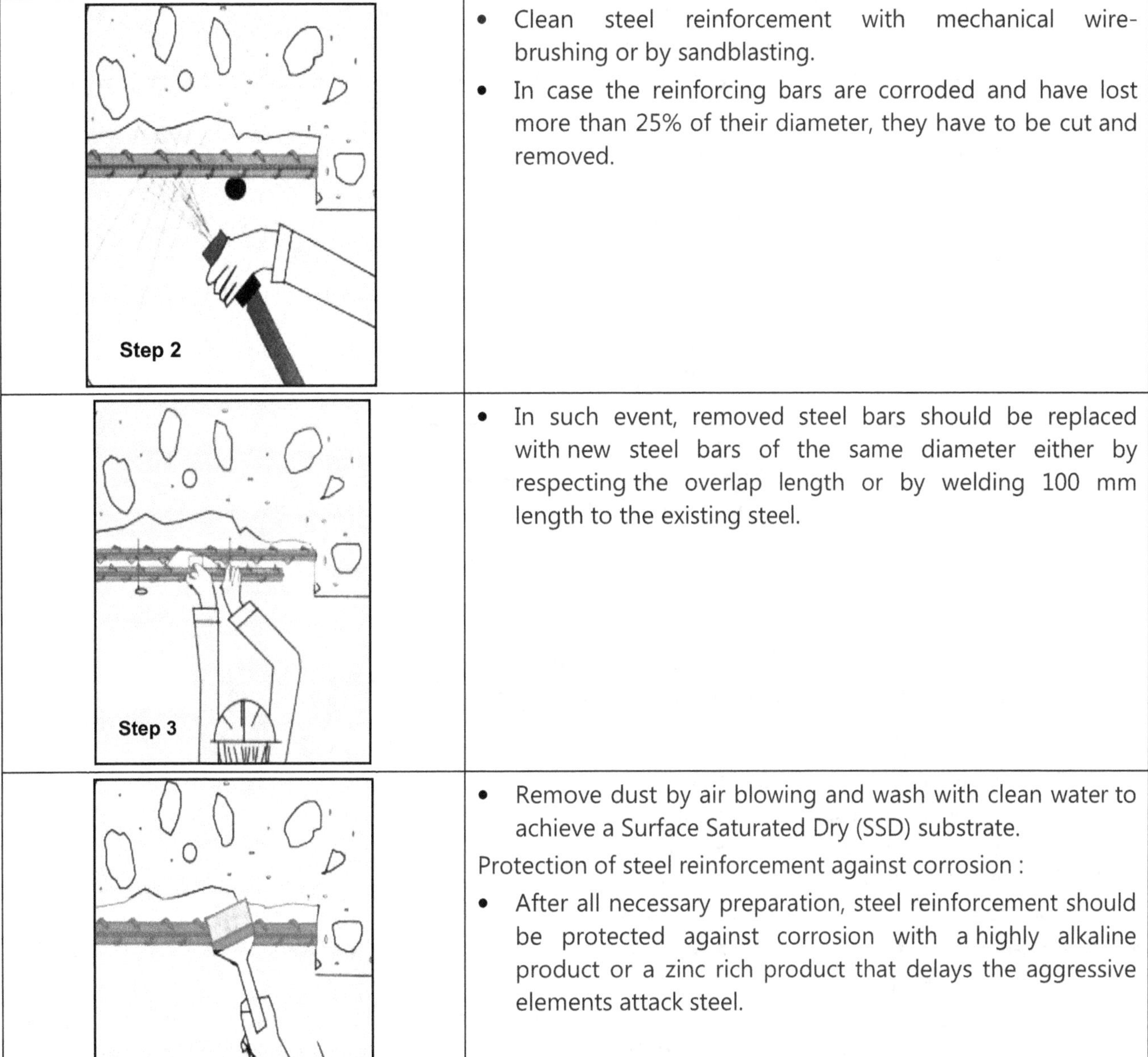

- Clean steel reinforcement with mechanical wire-brushing or by sandblasting.
- In case the reinforcing bars are corroded and have lost more than 25% of their diameter, they have to be cut and removed.

- In such event, removed steel bars should be replaced with new steel bars of the same diameter either by respecting the overlap length or by welding 100 mm length to the existing steel.

- Remove dust by air blowing and wash with clean water to achieve a Surface Saturated Dry (SSD) substrate.

Protection of steel reinforcement against corrosion :

- After all necessary preparation, steel reinforcement should be protected against corrosion with a highly alkaline product or a zinc rich product that delays the aggressive elements attack steel.

5.4.5 Provision of Additional Reinforcement

- Wherever the diameter of reinforcing bars is reduced substantially (say > 20%) additional bars shall be provided as per the design. This additional reinforcement shall be properly anchored to the existing concrete by providing adequate shear connectors. Weld mesh may also be provided if found necessary.
- Shear connectors of 8 mm diameter shall be provided in holes of 14 mm diameter and 75 mm deep. These shall be provided at every 500 mm c/c on all the faces of the beams in staggered form. The holes shall be cleaned with compressed air or water jet to remove all the dust etc. and then the shear connectors shall be fixed in the holes using polyester resin anchor grout.
- With the help of anchoring adhesives the deformed reinforcing bars can be installed in holes drilled in concrete to match the behavior of cast-in-place reinforcing bars. These are commonly referred to as post-installed reinforcing bars. This application can be characterized as follow :
  (i) Post-installed reinforcing bars are embedded in adhesive in a hole drilled in existing concrete on one side of the interface and are usually cast into new concrete on the other side of the interface. The bars may have hooks or heads on the cast-in one end, but are necessarily straight on the other end.

(ii) Post-installed reinforcing bars are often installed with small concrete cover ($3\phi > c > 2\phi$, where ϕ is the reinforcement bar diameter and c is the concrete cover). In such cases, the strength under tension loading of the post-installed reinforcing bar connection is typically limited by the splitting strength of the concrete.

(iii) Post-installed reinforcing bars are typically not designed to resist direct shear loading in the manner of an anchor bolt.

(iv) Post-installed reinforcing bars are generally embedded as required to "develop" their design stress using the basic required anchorage length, design anchorage length and splice length provisions. In order to achieve ductility of the structure the design stress will often be close to the design yield strength.

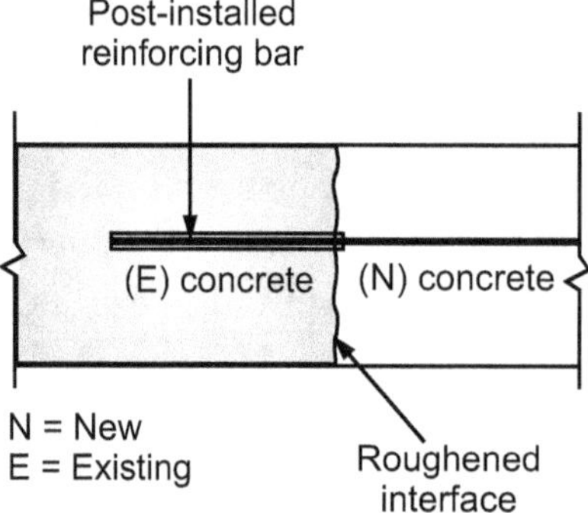

Fig. 5.13 : Additional reinforcement

5.5 STRENGTHENING METHODS FOR LIVE CRACKS

5.5.1 Repair Stages

1. Removal of damaged and loose concrete.
2. Cleaning of concrete surface with water/air jet.
3. Removal of all oil/grease from the surface.
4. Cleaning and sealing of cracks.
5. Roughening of surface to enhance bonding of repair materials.

5.5.1.1 Concrete Removal and Surface Preparation, (Various Methods of Surface Preparation)

- Expose the cracked/spalled elements completely. Use long sharp chisels of about 16-20 mm diameter and hammers up to 2 lbs weight. Remove the complete corrosion of reinforcement with wire brushes (preferably mechanical type). Remove all the loose and damaged concrete particles till sound concrete of uniform texture is visible.

- Apply rust removers to the reinforcement to remove the traces of rust. Clean the reinforcement once again with the wire brush. Wash the complete concrete surface including the reinforcement to remove the traces of rust remover.

- Fixing suitable formwork.

5.5.1.2 Bonding/Passivating Coat and Repair Application

- Apply a coat of 'rust passivator and cement' or equivalent as per the recommendations of the manufacturers. The concrete surface treated for corrosion shall be patch repaired with Polymer Mortar like as per the manufacturer's recommendations.

5.5.1.3 Separation Cracks at the Junction of Masonry and RCC Members

- These types of cracks are noticed almost all structures like sub-station building, office building etc. This is resulted due to dissimilar expansion of material i.e. concrete and masonry.

  Such cracks can be repaired as per the following steps :

  1. Expose the crack with sharp flat nose chisel and light weight hammers including removal of plaster approximately 6" inches on either side of cracks.
  2. Clean the cracks by vacuum cleaner.

3. Fill the crack with cement modified mortar i.e. 1 part of cement and 3 part of quartz sand with addition of nonshrink additive in the proportion as specified by the manufacturer.

4. Required quantity of water need to be added in the mortar to have desired consistency.

5. Application of bonding agent (Acrylic Polymer based bonding coats are added for advantage) in the proportion as specified by the manufacturer should be applied prior to application of cement modified mortar.

6. Replaster the area with the application of fiber mesh on the either side of crack.

5.5.1.4 Cracks Exist at the Junction of Subsequent Concrete Pours

- These types of cracks are observed particularly on the RCC wall construction horizontal junction between the subsequent pours as showing the cracks at many locations.

- These cracks are repaired as follows :

1. Open the cracks by making "V" grove with the help of mechanical cutters with least possible damage to the surrounding concrete areas.

2. Vacuum clean the cracks.

3. Seal the crack after application of polymer based bond coat with polymer modified mortar prepared by mixing of acrylic polymer cement and quartz sand in proportion of 1 : 5 : 15.

4. Applied mortar must be compacted at the laid position with the help of held vibrators. The application of the mortar shall be continued till the level of sealing material matches with the surrounding concrete surface.

5. Air cure polymer mortar as specified by manufacturer.

6. The cracks with surrounding concrete of suspect quality and width of crack is more than 2" inches i.e. 50 mm shall be filled with micro concrete with the application of bond coat as specified as per the manufacturer recommendations.

7. Structural cracks i.e. the cracks other than the separation, corrosion, junction, etc. exists in the failure zone of structural members. These type of cracks are identified in the data sheet of particular structures and have to be grouted with low viscous epoxy of standard manufacturer.

8. These cracks are also required to be opened up and to be sealed with epoxy putty. Low viscous epoxies are required to grout the cracks having width less than 0.2 mm with the pressure of 2.5 to 3.00 kg per sq. cm. Special nipples like brass nipples fixed properly along the crack with the help of epoxy putty/M-seal must be used to sustain such pressured injection.

5.5.1.5 Honey Combing, Weak Areas Around the Construction Joints

- These areas shall be grouted with cement grouting with addition of non-shrinking additives. Aluminum/PVC multi-perforated nipples may be used to carry out the injection operation. The pressure of 1.00 to 1.5 kg per sq. cm may be applied for grouting. Care should be taken to escape trapped air inside the crack; honey combing areas to avoid back pressure.

5.5.1.6 Severely Damaged RCC Members i.e. Columns, Beams, RCC Walls

- Members which are severely damaged in particular columns of pipe rack structure needs to be stabilized till the completion of entire rectification/remedial measures are carried out. These members shall be stabilized with epoxy mortar in proportion of 1 :6 to 1 :8 i.e. one part of epoxy and 6 or 8 part of quartz sand with the application of bonding coat of epoxy. Care must be taken to remove the dust around the application area with the help of vacuum cleaners as the entire efficacy of the epoxy is depending on the surface preparation.

5.1.1.7 Leakage, Seepage Through Expansion Joint

- Faulty expansion material in the joints must be replaced with the similar material with the improved material properties. Various options are available in the market. Area around joints must be closely inspected for the defects and should be rectified prior to the replacement of the joints. Injection grouting a week concrete areas also should be carried out whereever required.

5.1.1.8 Concrete Surface Exposed to Chemical Attack

- Concrete surface of various tanks exposed to severe chemical or chemical actions resulted due to various process, must be treated with anti-carbonation chemical coating. External surface of some of the tanks, exposed to chemical environment should also be treated with at the anticarbonation coating and painting.
- Different SMP's shall be available with Civil Department for improving reliability of equipment in safe condition. Audits for the same can be done for its implementation at site.
- Bonding aids :
  1. Cement slurry, polymers, epoxy.
  2. Polymer modified cement slurry.
- Anti-corrosive coating :
  1. Epoxies, zinc rich epoxy coating, bitumen.
  2. Fusion bonded epoxies interpenetrating polymer coating.

5.5.2 Repair/Strengthening Columns, Beams and Slabs

- These form the basic structural elements in most of the building structural systems, which are deteriorated and require attention to improve the load carrying capacity. Their structural modification or strengthening would give the required relief to the structure and enhance its performance as under :

Columns :

- The strengthening of columns may be required for the following :
  (a) Capacity : The load carrying capacity of the column can be enhanced by section enlargement. Different types of arrangement for section enlargement are shown in Fig. 5.17.
  (b) Ductility/confinement : The ductility of the column can be enhanced by providing additional tiles, steel plate bonding, and fibre wrap.
  (c) Joints : The joints play crucial for resisting earthquake forces. The joints can be strengthening by enlargement, jacketing by steel collar and fibre wrap.

Beams :

- These can be strengthened for :
  (a) Flexural Strength : The flexural strength of the beam can be enhanced by -
  1. Section enlargement in compression.
  2. Additional reinforcement in the tension. Caution shall be exercised to ensure that section is not over reinforced while providing additional reinforcement to compensate loss of reinforcement due to corrosion etc.
  3. The provisioning for enhanced tensile strength if being undertaken, this should be accompanied with corresponding increase in compression as well. Due to such increased flexural capacities extra shear capacities required to ensure ductile behaviour during earthquake shall also be considered for provision.
  4. MS plate bonding.
  5. High strength fibre fabric wrap technique (without section enlargement).
  (b) Shear Strength : The shear strength of the beam can be enhanced by any of the following :
  1. Section enlargement.
  2. Shear ties anchored in compression zone of beam.
  3. Post tension strap round the section.
  4. Diagonally anchored bolts (the holes are drilled perpendicular to the possible shear cracks).
  5. MS steel plate bonding.
  6. Fibre wraps.

Slabs :

- The performance of the slab can be improved by providing overlays (in case of negative moment deficiency) or underlay (in case of positive moment deficiency). The addition of overlay/underlay will also increase the stiffness of the slabs and control the excessive deflections problems. The slabs are generally safe in shear and as such no need is likely to occur for shear strengthening except flat slabs near column capital.

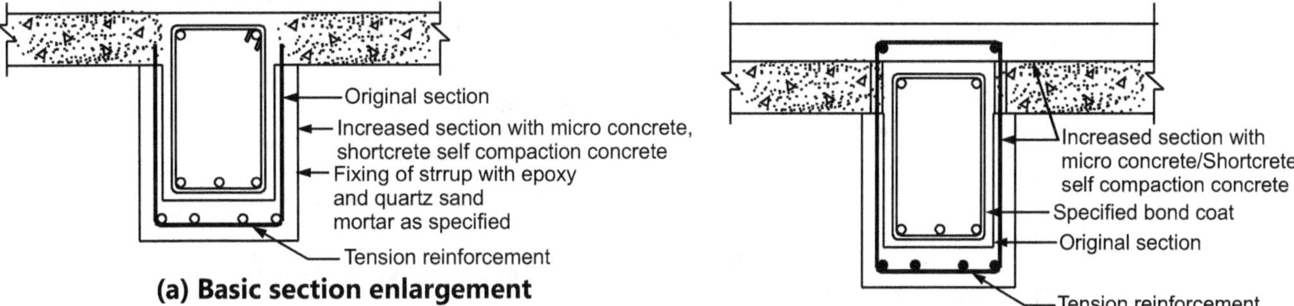

(a) Basic section enlargement

(b) Combination slab and beam overlay with beam enlargement

Fig. 5.14

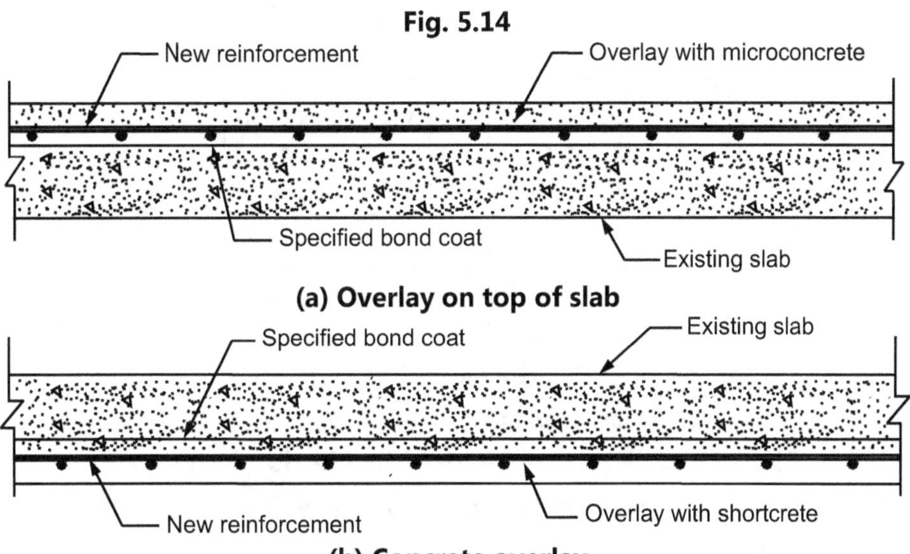

(a) Overlay on top of slab

(b) Concrete overlay

Fig. 5.15 : Slab strengthening

5.5.3 Crack Repair and Protective Coating for Less Damaged Structural Members Like Minor/Hair Cracks or Spalling from Beams and Columns, where Carbonation Depth in Cover Concrete has not Reached Reinforcement Level

- **Caution :** *It may be stated that recommended measures are cosmetic measures. The distress may continue to take place even after repairs, if the protective coating, provided as per step no. 7 below, is damaged or ineffective.*

Step-1 : Measures shall be taken to ensure that no seepage/leakage etc. affects the RCC columns/beams.

Step-2 : The plaster/finishes over the RCC columns/beams shall be removed carefully and the concrete surface is exposed. Spalled and loose concrete cover is removed, after close examination on the surface of concrete. The cracks are marked and the good surface of concrete shall be hacked and roughened for receiving the repair.

Step-3 : Wherever loose/spalled cover concrete is removed, it shall be repaired with polymer modified cement mortar, done up in layers.

Step-4 : All cracks in RCC columns/beams wherever noticed shall be sealed by injection grouting through nipples fixed along the crack line.

Step-5 : Over the prepared surface of RCC columns/beams, 6 mm thick 1 : 3 cement sand plaster shall be applied with polymer modified cement slurry bond coat within 24 hours of injection grouting.

Step-6 : Cement plaster shall be cured as per the relevant specifications.

Step-7 : After RCC columns/beams are cured and completely dried, a protective coating shall be applied over it for protecting the reinforcement and concrete against environmental aggressive chemicals.

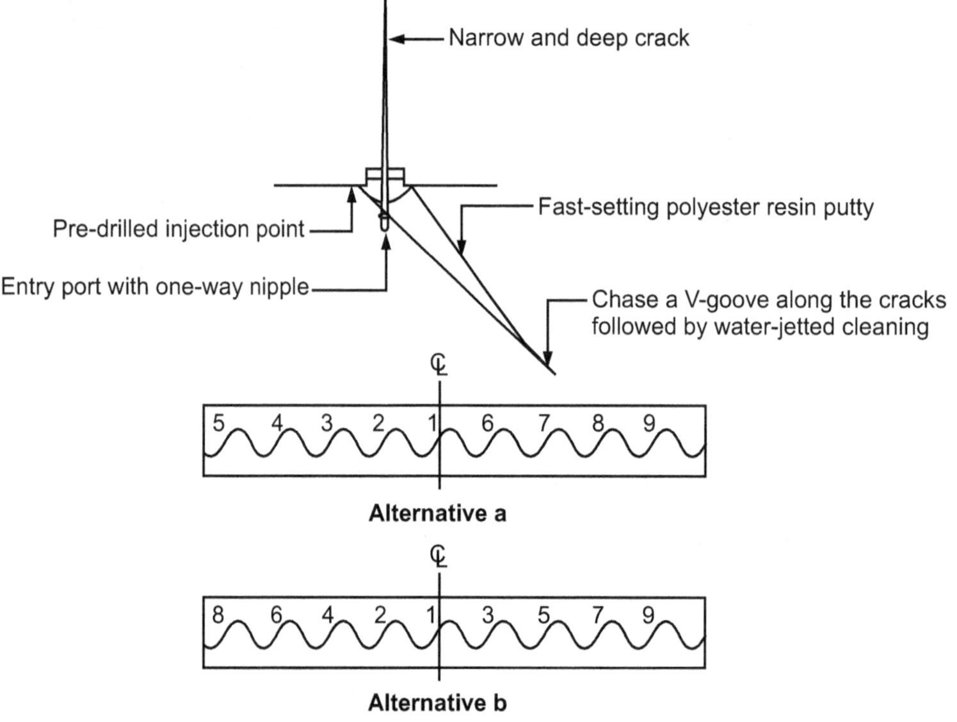

Fig. 5.16 : Crack sealing with resin injection technique

5.5.4 Repair to Damaged Columns/Beams where Carbonation Depth of Concrete has Reached Reinforcement Level

(i) Shotcreting :

Stepwise sequence of methodology to be adopted is given hereunder

Step-1 : Prop and support the structure in order to relieve the RCC column of stresses due to load coming over.

Step-2 : Remove plaster and finishes all around the distressed RCC columns. Thereafter remove loose, cracked and spalled concrete to expose the rusted reinforcement.

Step-3 : Remove concrete all around the reinforcement in order to get average 25 mm air gap all around the reinforcement and clean the reinforcement of concrete and rust by appropriate methods.

Step-4 : Put additional reinforcement wherever the reinforcement diameter has been reduced by more than 15-20% with the necessary overlap or welding with the existing reinforcement.

Step-5 : Fix shear key bars of appropriate diameter at specified spacings in both directions over the surface to be covered with repair materials.

Step-6 : Apply appropriate passivating and bond coat over the reinforcement and prepared RCC surface. Shotcrete the RCC column within the time limit specified as pot life of the epoxy or tacking period of slurry. The necessary shuttering as specified in specifications of shotcreting shall be used for ensuing the desired thickness and shape of the columns.

Step-7 : 6 mm thick finishing coat with cement sand plaster 1 : 3 (1 cement : 3 fine sand) (of least possible thickness) if felt necessary, shall be applied within 48 hours of application of shotcreted repair.

Step-8 : Wet curing shall be done over the finished surface of the shotcrete for a minimum period of 7 days.

Step-9 : After RCC columns/beams are cured and completely dried, a protective coating shall be applied over it for protecting the reinforcement and concrete.

(ii) RCC Jacketting :

Step no. 1 to 5 : Same as in (i) above.

Step-6 : Appropriate passivating and bond coat shall be applied over the prepared surface.

Step-7 : Within the tacky period of bond coat, shuttering and concreting shall be done with specified grade of concrete with minimum cement content as specified and water cement ratio not more than 0.45. The consistency of this concrete shall be flowing and self-compacting which shall be achieved by using super plasticiser. The thickness of RCC jacket shall be as specified.

Step-8 : Follow steps no. 6 to 8 as in (i) above.

For jacketing, micro concrete can also be used as per the requirement.

- **Formwork and shuttering :** Slurry tight and strong formwork shall be provided. The shuttering for encasement shall be kept ready such that the formwork shall be placed in position and fixed such that the micro concrete can be poured into the formwork within the overlay time of the bonding agent (5 hours). Adequate supports shall be provided for the formwork. Care should be taken to ensure leak proof shuttering. Under no circumstance the slurry should flow out of the shuttering during pouring of micro concrete.

- **Mixing of micro concrete :** It should be mixed using the appropriate water powder ratio as mentioned in the product data sheet. The mixing shall be done mechanically and under no circumstance hand mixing shall be done. Mixing shall be carried out for 3 to 5 minutes to ensure that homogeneous mix is obtained without any bleeding or segregation. In hot climate ice cooled water shall be used to maintain the temperature of mixed material. If the encasing thickness is more than 100 mm, add stone aggregates up to 50 % by weight of micro concrete to the mixed micro concrete directly into the mixer hopper. The stone aggregates must be 12 mm and down and shall be clean, washed and dried. The mixing should be done for 3 minutes in mixer and then pre-weighed stone aggregates into the mixer. Mix further for 2 minutes till lump free mix is obtained.

- **Pouring of micro concrete :** The mixer should be poured into the formwork using a suitable funnel or through a hose pipe. It must be poured from one end only. A suitable hopper/funnel arrangement shall be made at site to facilitate the pouring operations. The pouring operation shall be continuous and it shall not be stopped unless the job is completed. To achieve this sufficient mixers/drilling machines and wok force shall be arranged at site.

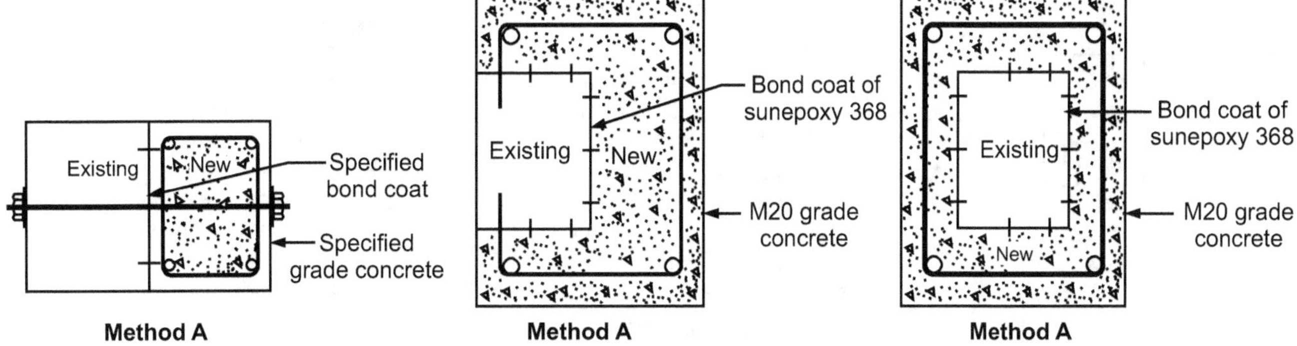

Fig. 5.17 : Column compressive strengthening by section enlargement

5.5.5 Stages for Repairs to RCC Slabs, where Carbonation Depth has Reached Reinforcement Level

(i) Repair with polymer modified cement mortar (For patch repairs or repairs over smaller magnitude) :

Stepwise sequence of methodology to be adopted is given below :

Step-1 : Propping and supporting of RCC slab under distress shall be done under the guidance of structural engineer.

Step-2 : All loose and spalled cover concrete shall be removed including finishing plaster wherever found loose by tapping.

Step-3 : The rusted reinforcement shall be cleaned of concrete preferably by using sand blasting to give a minimum 15 mm clear air gap all around including behind the reinforcement.

Step-4 : Additional reinforcement wherever necessary shall be added and tied to the RCC slab with necessary binding wires and nails.

Step-5 : Fix shear key bars of appropriate diameter at specified spacings in both directions over the surface to be covered with repair materials.

Step-6 : The rusted reinforcement shall be cleaned of rust and passivated and applied bond coating.

Step-7 : The prepared concrete surface shall be covered with appropriate mix of polymer modified cement sand mortar in layers including behind reinforcement over a bond coat with polymer modified cement slurry. The mortar cover thickness shall be not less than 15 mm over the reinforcement. The maximum thickness shall be not more than 30 mm with each layer not exceeding 10 mm.

(ii) Shotcreting : (For repair jobs covering large areas and/ or large magnitude).

Stepwise sequence of methodology to be adopted is given hereunder :

Step-1 to 6 : Shall be same as in case of no. (i) above.

Step-7 : Shotcreting with average thickness of 50 mm shall be done within the tacking period of epoxy bond coat to be applied over the prepared surface of concrete.

Step-8 : Finishing plaster if necessary, may be provided within 48 hours of shotcreting without allowing the RCC slab to become dry during the intervening period.

Step-9 : Water curing shall be carried out for a minimum period of 7 days.

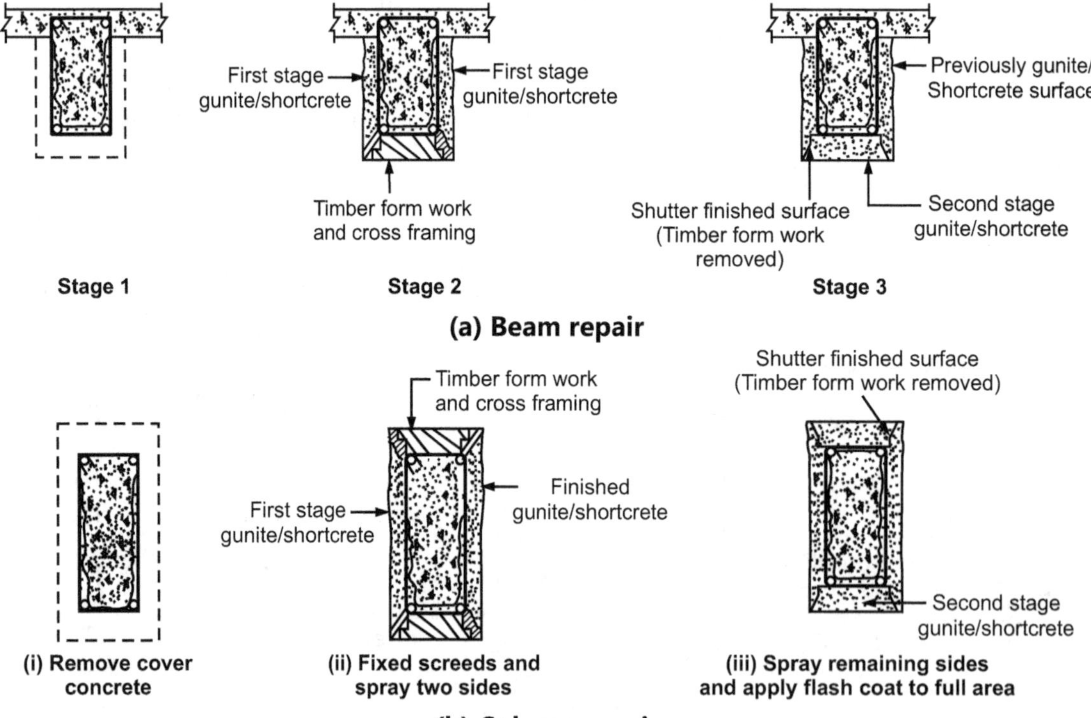

(a) Beam repair

(b) Column repair

Fig. 5.18 : Typical column by guniting

Important Points

- Concrete is a composite material that consists of embedded particles aggregates with a binding medium. In cement concrete, the binding medium is the mixture of cement and water. The concrete is said to be durable when changes occur at a rate, which does not adversely affect its performance within its intended life.

- Several types of cracks occur in RCC elements due to shear stress, corrosion of reinforcement, insufficient rebar cover, bending stress and compression failure etc.

- Physical deficiency is due to High w/c ratio and Inadequate curing, Poorly graded aggregates, Inadequate compaction, Shuttering joints not slurry tight, Cover thickness being lesser, Wrong placement of reinforcement, Heating/ Cooling, Wetting/ Drying.

- Chemical deficiency is due to chloride and sulphate attack. The reinforcement may get affected at an accelerated rate and initiate the onset of corrosion.

- When the roof slab is exposed to heavy showers of rain it becomes a major source of leakage in a structure particularly when it is not suitably protected. Leakages from roof slab and joints permit rainwater to enter in a structure.

- It is essential to provide adequate slope to the flat and sloping roof to ensure effective drainage. The slope of roof should be such that the water does not accumulate and drained off quickly under gravity. A slope of 1 in 100 or steeper, depending upon the type of water proofing system, is required to drain off water effectively.

- Epoxy injection of a low viscosity epoxy is a possible repair method for cracks. The epoxy injection to be effective the crack must be free of dirt grease or other contaminations.

- Dry pack is suitable for filling holes whose depth is at least equal to the smallest surface dimension of the repair area. The holes should be at least 25 mm deep. Dry pack is not suitable for shallow depressions. The holes that go right through concrete section where the filling cannot be properly rammed.

- Grooving and sealing is the simplest and most common method of crack repair. It can be executed with relatively unskilled personnel and can be used to seal both fine pattern cracks and larger isolated cracks.

- In stitching the crack is bridged with U-shaped metal units stitching dogs before being repaired with a rigid resin material. This can establish restoration of the strength and integrity of cracked section; due care is to be given to make analysis check to ensure that this will perform well under applied loads. A non-shrink or an epoxy resin based adhesive should be used to anchor the legs of the dogs.

- Bandaging a flexible strip is fixed over the crack with only the edged of the strip bonded. Where movement is not all in one plane, where is excessive movement beyond that which can be accommodated by a recess of convenient size, or if there are factors which prohibit the cutting of a recess, a surface bandage can be used. In areas which are subject to traffic, the flexible bondage will be coated over with a wearing course.

- When the cement-water/chemical comes out through the adjacent holes the injection of cement-water/chemical is stopped and the plugs in the bottom most hole and the one immediately above are restored. The process of grouting of concrete cracks is repeated with other holes till all the holes are covered. On the day of grouting all the plugs are removed to drain out excess cement-water/chemical and restored before commencing grouting.

- Gunite can be defined as, "mortar conveyed through a hose pneumatically protect at a high velocity on to a surface". The method has been further developed by the introducing of small size coarse aggregate into mix. This process economical by reducing the cement content.

- The fibre wrap technique, also known as Composite Fiber System is a non-intrusive structural strengthening technique that increases the load carrying capacity and ductility of reinforced concrete members without causing any destruction or distress to the existing concrete.

- Corrosion is defined as, "the process of deterioration (or destruction) and consequent loss of a solid metallic material, through an unwanted (or unintentional) chemical or electro-chemical attack by its environment, starting at its surface.". Thus, corrosion is a process of "reverse of extraction of metals".

- A corrosion inhibitor is an admixture that is used in concrete to prevent the metal, embedded in concrete form cording. There exists various types of inhibitors like cathode, anode, mixed and dangerous, safe.

- Cathode protection is one of the effective, well known, and extensively used methods for prevention of corrosion in concrete structures in more advanced countries. Due to high cost and long term monitoring required for this method, it is not very much used in India.

- Galvanized Reinforcement : Galvanizing of reinforcement consists of dipping the steel bars in molten zinc. This results in a coating of zinc, bonded to the surface of steel. The zinc surface with calcium hydroxide in the concrete, to form a passive layer and prevents corrosion.

Questions for Practice

1. Explain the steps in repair aspect.
2. Define shotcrete. Explain the two types of process in Shotcrete.
3. Define Gunite.
4. Explain the stages in dry mix process in shotcrete.
5. Describe the steps in the assessment procedure for evaluate damages in a structure.
6. Explain the various causes for deterioration of concrete structures.
7. Define corrosion inhibitor. Give some examples for corrosion inhibitors.
8. Explain chloride attack on concrete?
9. Explain in detail about expansive cement.
10. Briefly explain about polymer concrete and its types.
11. Explain in detail about Sulphur infiltrated concrete.
12. Explain techniques required for repairing cracks.
13. Define stitching. Define external stressing.
14. Explain blanketing.
15. Give short note on Jacketing.
16. Define grouting.
17. Give a brief account on routing and sealing.

...

Chapter **6**

STRUCTURAL AUDIT AND BUDGET

Weightage of Marks = 12, Teaching Hours = 12

Syllabus

6.1 Necessity and importance of structural audit and budget estimation.

6.2 Distress survey, detailed inspection, recommendations for budget estimation.

6.3 Steps involved in structural audit and budget estimation.

6.4 Format preparation for structural audit including general information of building, building data, complain reported by users, inspection of internal and external areas of building.

6.5 Overview on rules and regulations of structural audit and budget estimation as recommended by competent authority such as Public Work Department.

Objectives

After learning this chapter, student will be able to

- Explain the necessity and importance of structural audit and budget estimation.

- Explain the procedure involved in structural audit and budget estimation.

- Explain the step-by-step procedure for maintenance of the given structure.

- Know the formats preparation for the process of structural audit and budget preparation.

- Explain the rules and regulations of structural audit and budget estimation as recommended by competent authority.

INTRODUCTION

- In India there are many old buildings, which have reduced the strength in due time. If further use of such depreciation structure is continued, it can endanger the lives of occupants and the surrounding habitat. There must be appropriate action should then be implemented to improve the performance of structures and restore the desired function of structures. Thus, it is extremely important to do a structural audit of the existing building and implementation of maintenance/repair work lead to sustained life of building and safety of the resident.

- To be more responsible and attentive towards old buildings, municipal corporation has to issue notice to buildings and co-operative societies which are more than 30 years old to carry out mandatory structural audit and submit audit report. Structural audit should highlight and investigate all important areas and make immediate remedial suggestions and preventive measures. It should cover structural analysis of existing frames and find key elements for all types of loading. It also helps to deliver a stronger building structure with cost-effective solutions and proper maintenance program.

- This chapter deals with the study of various parameters of structural audit including visualization inspection, non-destructive testing, core sampling and testing. It is also emphasized various repair and retrofitting measures used for buildings after structural audits.

(6.1)

6.1 STRUCTURAL AUDIT

- Structural Audit is an overall health and performance check-up of a building like a doctor examines a patient. It ensures that, the building and its premises are safe and have no risk. It analyses and suggests appropriate repairs and retrofitting measures required for the buildings to perform better in its service life. Structural audit is done by an experienced and licensed structural consultant.

- "Structural audit is the inspection or examination of the building, to evaluate the strength so as to improve its appropriateness, safety, efficiency". This extent of damage or deterioration depends on the quality of work at the construction level. The deterioration of buildings can be the result of various factors including fire damage, frost action, chemical attack, steel erosion during the life span of the structure.

6.1.1 Purpose of Structural Audit

- To save human life and buildings.
- To understand the condition of building.
- To find critical areas for immediate repair.
- To comply with municipal or any statutory requirements.
- To enhance life cycle of building by suggesting preventive and corrective measure such as repairs.
- To know the health of building and to improve the expected future life.
- To highlight the critical areas that need to be attended with immediate effect.

6.1.2 Importance of Structural Audit

- To enhance the overall lifecycle of the building, its component need to be periodically examined so that there is no danger to its residents.
- Areas in need of crucial repairs need to be identified to make use of corrective measures.
- The day to day life of the residents is not inconvenienced on account of unexpected leakages.
- A structural audit is highly recommended preventive measure to avoid collapse during monsoons or due to ill-maintenance.
- To prevent any disasters, in the probable circumstances that the builder has left, any sections of a building prone to collapse a structural audit plays important role.
- The only way to avoid loss of human life in addition to property is to conduct structural audit with highest sincerity and apply remedial measures immediately.
- For a sustainable and safe community of mindful residents structural audit is important timely.

6.2 DISTRESS SURVEY

- Distress survey is an examination of concrete for the purpose of identifying and defining area of distress. While it is referred in connection with survey of concrete and embedded reinforcement that is showing some degree of distress, its application is recommended for all buildings and structures. The system is designed to be used for recording the history of the project from its inception to completion and subsequent life.

6.2.1 Objectives

- The objectives of Condition Survey of a building structure are :
  - (a) To identify
    - causes of distress, and
    - their sources;
  - (b) To assess -
    - the extent of distress occurred due to corrosion, fire, earthquake or any other reason,
    - the residual strength of the structure, and
    - its rehabilitability;
  - (c) To prioritise the distressed elements according to seriousness for repairs; and
  - (d) To select and plan the effective remedy.

6.2.2 Stages

- Stages for carrying out condition survey, largely depend on field conditions, user habits, maintenance etc. and have a direct relation with the pattern of distress, whether localized or spread over. Condition Survey of a building/structure is generally undertaken in four different stages to identify the actual problem so as to ensure that a fruitful outcome is achieved with minimum efforts and at the least cost. The four stages of condition survey described in Fig. 6.1 are :

(a) Preliminary inspection.

(b) Planning.

(c) Visual inspection.

(d) Field and laboratory testing.

I.   Preliminary Inspection :

- The primary objective of the preliminary inspection is to assess and collect following necessary information for a thoughtful planning before a condition survey is physically undertaken.

1. Background history of the distressed structure :

    –   from the Owners/Clients;

    –   from the occupants of building, general public, etc. based on personal enquiries.

2. Notes and records of earlier repairs, if carried out.

3. All possible relevant data and information.

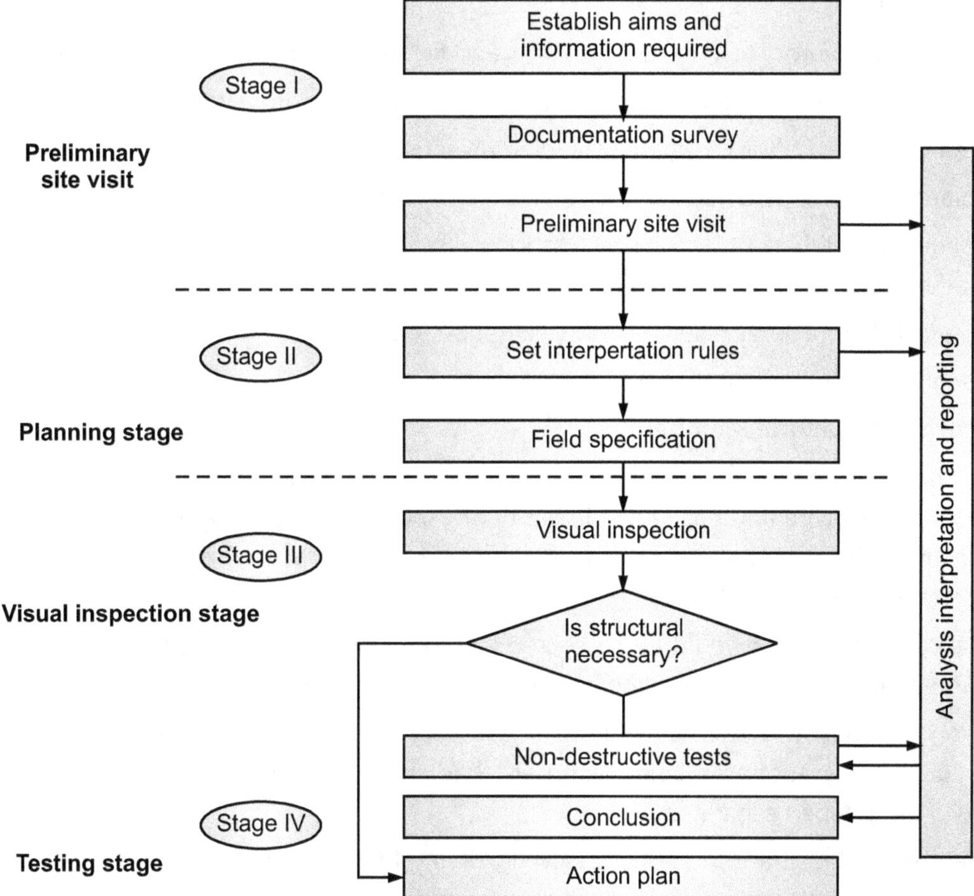

Fig. 6.1 : Flowchart to illustrate the process of condition survey

4. The practical restrictions in conducting field survey and devise methods to overcome the same;

5. The safety requirements for condition survey team;

6. Necessary site preparations including access scaffolds, working platforms, etc, if any;

7. The extent and quantum of survey work;

8. The approximate time required for survey;

9. (i) The requirement of

      - field-testing equipments,

      - tools for sampling.

    (ii) To advise the client/owner of the building in regard to immediate safety measures, if considered necessary, to avert any mishap endangering life and structure.

    (iii) To define the scope of work of field investigations in consultation with the Clients/Owners.

II. Planning Stage :

- Planning stage involves preparation of field documents, grouping of structural members and classification of damage as under:

(i) Preparation of Field Documents :

- For distress survey, the following are required to be prepared:

(a) Survey objective.

(b) Scope of work.

(c) Method of survey.

(d) The field and laboratory testing requirements and field equipments and tools required for the same.

(e) List of tasks and their sequence for condition survey together with a work schedule.

(f) Required number of photo copy of available drawings.

(g) Floor plans based on field measurements.

(h) Work sheets and tables for recording in a logical manner all information, test results including field data gathered.

(i) Previous Condition Survey results and Investigation Reports, if any.

(j) Maintenance and repair records.

(ii) Grouping of the Structural Members :

- Soon after the preliminary site visit and on perusal of building plans, the structural members shall be grouped as per their type and based on similarity of exposure conditions for proper appreciation of the cause of distress. For example, in a building subjected to normal environmental attack, the grouping could be done as under :

(a) External columns/beams would be subjected to more severe environmental attack than the internal structural members of a building and could be grouped in two broad groups.

(b) Even from amongst the external columns, those at corners or projected out are likely to be exposed more due to adjacent faces being exposed than those not at corners or un-projected columns. Hence, to be grouped separately.

(c) The members subjected to dampness/wetting/drying located in or around the toilet shafts are likely to undergo similar class of distress and be grouped separately.

(d) Structural members with different protective finishes have to be grouped separately.

(a) Classification of Damage :

- Based on the preliminary data collected and site visit, the rehabilitation engineer should freeze the interpretation rules and subdivide the repair classification broadly in to five classes as 'Class 0' to 'Class 4' named as Cosmetic Repair, Superficial Repair, Patch Repair, Principal Repair and Major Repair.

- The classification given in Table 6.1, is generally considered sufficient for deciding the 'Repair Requirements' for carbonation induced corrosion damaged structure. This is just an indicative classification and the same may vary from case to case. It should be decided by the Rehabilitation Engineer before taking up detailed Condition Survey of the distressed structure in field.

(b) Visual Inspection :

- Visual Inspection of a structure is the most effective qualitative method of evaluation of structural soundness and identifying the typical distress symptoms together with the associated problems.

- This provides valuable information to an experienced engineer in regard to its workmanship, structural serviceability and material deterioration mechanism.

- It is meant to give a quick scan of the structure to assess its state of general health.

- The record of visual inspection is an essential requirement for preparation of realistic bill of quantities of various repair items.

- Experienced engineers should carry out this work as this forms the basis for detailing out the plan of action to complete the diagnosis of problems and to quantify the extent of distress.

- Simple tools and instruments like camera with flash, magnifying glass, binoculars, gauge for crack width measurement, chisel and hammer are usually needed. Occasionally, a light platform/scaffold tower can be used for access to advantage.

(c) Field/Laboratory Testing Stage :

- It may neither be feasible nor is the practice to conduct field/laboratory testing on every structural member in an existing distressed building.

- The field/laboratory testing of structural concrete and reinforcement is to be undertaken, basically for validating the findings of visual inspection.

- These may be undertaken on selective basis on representative structural members from each of the various groups based on exposure conditions as explained in the preceding sections.

- The programme of such testing has to be chalked out based on the record of visual inspection.

Notations for Symptoms of Distress :

Table 6.1

Sr. No.	Symptom of Distress	Notation
1.	Loose cover concrete	
2.	Spalled cover concrete with reinforcement not exposed	
3.	Spalled cover concrete with reinforcement exposed	
4.	Crack in RCC slab	b = length of crack
5.	Crazing cracks	Crazing

Table 6.2

Sr. No.	Symptom of Distress	Diagram	Notation
1.	Loose cover concrete.	1250-500 (a-b)	(a) - Distance flow left face of column. (b) - Length of loose cover concrete
2.	Crank along reinforcement in the soffit of beam.	1250-2000 (a-b-c)	(a) - Starting point of crack from left face of on column/wall. (b) - Length of crack. (c) - Crack width in mm.
3.	Diagonal or vertical crack on vertical face of a beam (crack to be shown inclined or vertical as the case may be).	W 1250-250 (a-b-c)	(a) - Starting point of crack from left face of column/wall. (b) - Length of crack. (c) - Crack width in mm.
4.	Spalled cover concrete with no exposure of reinforcement.	1250-300 (a-b)	(a) - Distance from left face of column. (b) - Length of loose cover concrete.
5.	Spalled cover concrete with reinforcement exposed.	1250-500 (a-b)	(a) - Distance from left face of column. (b) - Length of loose cover concrete.

6.3 STAGES IN CARRYING OUT STRUCTURAL AUDIT

Stages in Structural Audit :

- Study of architectural and structural drawings, design criteria, design calculations, structural stability certificate of existing structures.
- If architectural plans and structural plans are not available, it can be prepared by any engineer.
- Inspection of the building.
- Preparation of audit report.

6.3.1 Step by Step Procedure to be followed in Structural Auditing

- **STEP 1 :** It is imperative that we must have Architectural and Structural plans of the buildings. It will be helpful if we have detailed structural calculations including assumptions for the structural design.
- **STEP 2 :** If the Architectural plans and Structural plans are not available, the same can be prepared by any Engineer.
- **STEP 3 :** Inspection of the building - A detailed inspection of the building can reveal the following :
    1. Any settlements in the foundations.
    2. Cracks in columns, beams and slabs.
    3. Concrete disintegration and exposed steel reinforcement photographs can be helpful.
    4. Slight tapping using hammer can reveal deterioration in concrete.

5. Corrosion in reinforcement.
6. Status of Balconies – sagging, deflection, cracks.
7. Status of Architectural features viz. Chhajjas.
8. Cracks in walls indicating swelling in R.C.C. members or deflection or corrosion.
9. Leakages from terrace and toilet blocks.
10. Leakages and dampness in walls resulting into cracks and corrosion.
11. Status of repairs and last repaired.
12. What was repaired?
13. Who was the Agency?
14. How much was spent for repairs?
15. Building plans are available? When approved?

- **STEP 4 :** Preparation of Audit Report : On the basis of inspection of building an Audit Report is prepared.
- **STEP 5 :** Tests Recommended : It is important that various tests are carried out in the old buildings. This will give an idea about the extent of corrosion, distress and loss of strength in concrete and steel.
  STEP 6 : Highlight the critical areas and how to go for repairs.

I. Visual Inspection :

(a) Need of visual inspection :

- To identify the types of structural defects.
- To identify any signs of material deterioration.
- To identify any signs of structural distress and deformation.
- To identify any alteration and addition in the structure, misuse which may result in overloading.

(b) Points to be taken under consideration while visual inspection :

- The inspection report should reveal the following along with photographs and sketches :
  - Settlement of columns or foundations.
  - Settlement of walls and floors.
  - Deflection and cracks in retaining wall.
  - Materials used and framing system of structure.
  - Identification of the critical structural members like floating columns, transfer beams, slender members, rusting of exposed steel and its extent.
  - Status of all building elements like beams, slabs, columns, balconies, canopy, false ceiling, chajja, parapet and railings with respect to parameters deflection, cracks, leakages and spalling of concrete.
  - Status of water tank, staircase, lift and lift machine room.
  - Dampness in walls.
  - Leakages in terrace, toilets, plumbing lines, drainage lines and overhead tanks.
  - Blistering of paints and paint peel off.
  - Inspection of drainage system.

Fig. 6.2 : Exposed reinforcement in wall

Fig. 6.3 : Cracks in beams

Fig. 6.4 : Exposed brickwork in walls

II. Scope of Visual Inspection :

- The inspection report should reveal the following listings along with photographs and sketches :

(a) General information of the building :

- Name and address of the building.
- Number of stories in each block of building.
- Description of main usage of building viz. Residential, commercial, institutional.
- Maintenance history of the building.

(b) Structural system of the building :

- Sub-structure: Settlement of columns or foundations, disposal of walls and floors, deflection and cracks in retaining walls, soil bearing capacity through trial pits or from adjacent soil data.
- Super structure: Framing system structure and material used, identification of critical structural members such as floating columns, transfer beams, tapered members, exposed steel corrosion and its extent.
- Mention the condition of all construction elements such as beams, slabs, columns, balconies, canopy, false ceilings, chajja, parapets, and railings with parameters deflection, cracks, leakage and swallowing of concrete.
- Similarly, verify the status of the water tank, staircase, lift and lift machine room.

(c) Addition or Alterations in the building :

- Identification of change of occupancy.
- Alteration or addition of partition walls.
- Alteration or addition in loadings - stacking.
- Alteration or addition of toilets, water tank.
- Alteration or addition of balcony.

(d) Dampness and leakages :

- Detect the dampness in walls.
- Identify the leakages in terrace, toilets, plumbing lines, drainage lines and overhead tanks.

6.3.2 Destructive and Non-Destructive Testing

In addition to visual inspection next stage to structural audit is Non-Destructive and Destructive Testing.

Destructive Testing :

- To verify the integrity of a component, it is always possible to cut or section through the components and examine the exposed surfaces. Components can be pulled or stressed and pressurized until failure to determine their properties of strength and toughness. Materials can be chemically treated to determine their composition. These are some forms of destructive testing. Unfortunately this approach of destructive testing renders the component useless for its intended use as against non-destructive testing that can be performed on the components and machines without affecting their service performance.

Non-Destructive Testing :

- Non-destructive testing (NDT) is a broad set of analysis techniques used in science and the technology industry to evaluate the properties of a material, component, or system without damage. The terms non-destructive examination, non-destructive inspection and non-destructive evaluation are also commonly used to describe this technique, as NDT does not permanently damage the structural member being inspected. It is a highly valuable technique that can save both money and time in building evaluation. There are several non-destructive tests (NDTs) exist for concrete members that determine the present strength and quality of concrete. Some of these tests are very useful in assessing damage to RCC structures due to corrosion, chemical attack, fire and other causes.
- These tests have been put under four categories depending on the purpose of test as under :

(A) Concrete Strength :

- **Rebound Hammer Test :** To measure surface hardness of concrete.
- **Ultrasonic Pulse Velocity Test :** To assess homogeneity of concrete, to assess strength of concrete qualitatively, to determine structural integrity.
- **Core Sampling and Testing :** To measure strength, permeability, density of concrete.

(B) Chemical Attack :

- **Carbonation Test :** To assess depth of carbonation and pH of concrete.
- **Chloride Test :** To assess total water/acid soluble chloride contents.
- **Sulphate Test :** To assess total water/water soluble sulphate contents of concrete.

(C) Corrosion Potential Assessment :

1. **Cover Meter :** To measure cover of reinforcement, diameter of reinforcement and spacing of reinforcement.
2. **Half Cell Method :** To assess probability of corrosion in the embedded steel.
3. **Permeability Test :** To assess permeability of concrete due to water and air.

- **Following Non-destructive tests are widely practiced to determine the present strength and quality of structure :**
  1. Rebound Hammer Test.
  2. Pulse Echo Method.
  3. Impact Echo Method.
  4. Ultra Sonic Pulse Velocity Method.
  5. Probe Penetration Test or Windsor Probe Test.
  6. Ground Penetration Radar Method.
  7. Carbonation Test.
  8. Half Cell Potential Meter Test.

6.3.3 Identification of Critical Areas in Building

- Based on the above inspection, analysis and test results, the report should conclude the critical areas that need immediate repairs and retrofitting. For example: number of columns requiring immediate repair and strengthening, repair of critical slab and beams, water proofing of terrace, toilet blocks, cracks in walls or structural elements etc.

6.3.4 Post Structural Audit

(A) Repairs :

- Based on the audit findings and recommendations different measures of repairs and strengthening are carried out. According to the report, repair will include replacing or correcting of deteriorated, damaged, or faulty materials, components, or elements of a structural system. From this point of view, repair may be divided into structural repair and serviceability repair. The former refers to the restoration of lost sectional or monolithic properties of damaged members, while the later refers to the restoration of structural surfaces to a satisfactory operational standard.

- Obviously, poor design, poor construction, poor maintenance, incorrect usage, new environmental influences or an intended increase of the loading or extension of the structure's lifespan can make repair and/or strengthening necessary. Excluding technical considerations, the ultimate choice of method of repair and strengthening of a concrete structure may also be influenced by factors like overall quality of repairs and the size of individual repairs, access for repair, relative cost, ease of application, available labor skills and equipment and client requirements including future maintenance and economic considerations.

(B) Strengthening and Retrofitting :

- Strengthening is the process of restoring the capacity of damaged components of structural concrete to its original design capacity, or increasing the strength of structural concrete.

- Strengthening of a concrete structure may be required due to several reasons:

    - Change of usage which may cause over-stress in the structural member.

    - Serious materials and structural deteriorations which cause structural members to be no longer able to carry the imposed loads with an adequate factor of safety.

    - To increase the capacity for seismic resistance if the building is not designed for it or the structure does not fulfill current design requirement corresponding to seismic zones, R factor or so. Strengthening of structural members can be achieved by replacing poor quality or defective material with better quality material, by attaching additional load-bearing material, such as high quality concrete, additional steel, thin steel plates, various types of fiber reinforced polymer sheets, and so on, and by the redistribution of the load such as by adding a steel supporting system.

    - To increase the load-carrying capacity or stability of a structure with respect to its previous condition.

6.4 AUDIT REPORT

- An Audit Report is prepared on the basis of the inspection carried out.
- General Format of the Structural Audit Report :
    1. Name of the building.
    2. Name of the owner.
    3. Address.
    4. Contact number.
    5. Year of construction.
    6. Name of structural engineer for audits.

Table 6.3 : General Observation

Sr. No.	Description	Remark
1.	Type of building structure	
2.	Age of building	
3.	Number of wing	
4.	Mode of use	
5.	Number of storeys	
6.	Number of tenants	
7.	Plan available : Architecture/Structural	

Table 6.4 : Structural Observation

Sr. No.	Description	Component	Grade
1.	Cracks	Beam	
		Column	
		Slab	
		Plaster	
		Wall	
2.	Settlement	Foundation	
		Joint at plinth	
		Column	
		Wall	
3.	Leakage and Dampness	External wall	
		Toilet	
		Terrace	
		Slab	
		Water tank	
		Drainage line/Pumping line	
4.	Deflection	Beam	
		Slab	
		Balcony	
5.	Condition of staircase, balcony, flooring and ducts		

RCC framed structures are rated by grades as follows :

Table 6.5

Sr. No.	Grade	Description	Colour Code
1.	0 to 3	Major distress	Red
2.	3 to 5	Considerable distress and repairable	Yellow
3.	5 to 7	Modrate distress and repairable	Blue
4.	7 to 10	Sound structure	Green

Table 6.6 : Observations during Inspection of Common Areas

1.	**External Walls**			
	External Face			
	Chhajja			
	Beam - Wall			
	Wall - Column			
	Toilet Area			
2.	**Staircase**	A Wing	B Wing	C Wing
	Walls			
	Jali			
	Flooring			
	Stair			
	Plaster			
	Beams			
3.	**Water Tank**			
	Pipes Joints			
	waterproofing			
	Walls			
4.	**Plumbing**			
	Pipes			
	Joints			
	Walls junctions			
5.	**Sanitary Lines**			
	Pipes			
	Joints			
	Chambers			
6.	**Terrace**			
	Slab			
	Water Proofing			
	Flooring			
	Coping			
	Parapets			
	RWP			

7.	**Compound Area**	
	UGT	
	Sewer Line	
	Chambers	
	Rooms	

	P R O F O R M A	
	Name of Consultant :	
1.	Name of Building	
2.	CTS No./Ward	
3.	No. of Storey	
4.	Year of Construction	
5.	User Department	
6.	**Mode of Construction of Existing Building :**	
	(i) Type of foundation	
	(ii) Type of floors	
	(iii) Type of walls	
	(iv) Type of beams	
	(v) Type of columns	
	(vi) Type of roof	
7.	**History of Repairs done year wise :**	
	(a) Slab recasting	
	(b) Column jacketing	
	(i) Structural repairs	
	(ii) Tenantable repairs	
	(iii) Roof/Waterproofing	
	(iv) Plumbing	
	(v) Additions/Alteration if any	
8.	**Date of Inspection by Consultant :**	
9.	**Condition of :**	
	(i) Internal plaster	
	(ii) External plaster	
	(iii) Plumbing	
	(iv) Drains lines/Chambers	

10.	**Observations :**		
	(a) Doors and Windows (Operational/Non-operational)		
	(b) Columns and Steel exposed		
	(c) Settlement uneven flooring gaps between skirting and floor		
	(d) Foundation settlement		
	(e) Deflection/Sagging		
	(f) Major cracks in column/beams		
	(g) Seepages/Leakages		
	(h) Staircase area/Column condition		
	(i) Lift walls		
	(j) U. G. tank		
	(k) OHT/Column condition		
	(l) Parapet at terraces		
	(m) Chhajjas		
	(n) Common areas		
	(o) Toilet blocks		
	(p) Terrace/Waterproofing		
11.	**Test carried out on structure/observations thereof :**	**Finding**	**Range as per IS Code**
	(a) Ultrasonic pulse velocity test		
	(b) Rebound hammer test		
	(c) Half cell potential test		
	(d) Carbonation depth test		
	(e) Core test		
	(f) Chemical analysis		
	(g) Cement aggregate ratio		
12.	**Distress mapping plan and photographs with caption below about description of structural member and its location**		

13.	**Brief Description of Repairs to be Done**		
	(a) Water proofing		
	(b) External plaster		
	(c) Structural repairs		
	(i) Column jacketing		
	(ii) Slab recasting		
	(iii) RCC cover to be replaced		
	(iv) Beam recasting		
	(d) Partial evacuation during repairs needed		
	(e) Propping		
14.	**Conclusions of consultants**	**Observations /Key reason**	
I.	Whether structure is livable/or whether it is to be evacuated/pulled down.		
II.	Whether structure requires tenantable repairs/ major structural repairs and its time frame.		
III.	Whether structure can be allowed to occupy during course of repair.		
IV.	Nature/Methodology of repairs.		
V.	Whether structure requires immediate propping. If so, its propping plan/ methodology given.		
VI.	Whether other immediate safety measures required- what is specific recommendation?		
VII.	Enhancement in life of structure after repairs/frequency of repairs required in the extended life period.		
VIII.	Projected repair cost /Sq.ft.		
IX.	Projected reconstruction cost /Sq.ft.		
X.	Specific remarks, whether building needs to be vacated/demolished/repairable.		
XI.	Whether structure in extremely critical condition.		
15.	**Critical observation**		
16.	**Classification of buildings**	Category	Auditors final conclusion

6.5 OVERVIEW ON RULES AND REGULATIONS OF STRUCTURAL AUDIT

- As per Model Bye-Laws for Co-operative Housing Societies under the Maharashtra State Co-operative Societies Act, duly amended (upload date 8/10/2014 (version dated 2-9-2014) Bye law no 75 (Society to carry out Structural Audit),

  (a) The Society shall cause to undertake the Structural Audit of the building as follows :

    (i) for buildings aged between 15 to 30 years … once in 5 years.

    (ii) for buildings aged more than 30 years … once in 3 years.

  (b) Such Structural Audit by Societies which are in Municipal Corporations limits shall be conducted by approved Engineers from the Corporations panel. In case of other Societies such structural audit shall be carried by the Govt. Approved Structural Engineers/Architect, and maintain record thereof.

  (c) The Society shall undertake to carry out periodical Fire Audit of its property as per the State Fire Policy, and maintain record thereof.

  (d) The Society shall carry out periodical Inspection of Lifts/Elevators and maintain record.

  (e) Taking into consideration all factors such as negligence in maintenance and illegal alteration which include shifting of WC, bath and kitchen etc., the local civic bodies, lawfully made it mandatory for the building owners/Societies to conduct "Structural Audit", for all those buildings which are over 20 year of age.

  (f) The Structural Engineer, mandatorily must be a NEUTRAL authority, more so since any manipulations or fabricated or untrue Structural Audit Report, would mean, criminal prosecution against the Structural Engineer.

  (g) It shall be mandatory for the Municipal Authorities, to initiate other legal proceedings, against the building Owners/Society, in the absence of Approved /Sanctioned Plans, Occupancy Certificate and other documents. This ultimately serves to safeguard the residents of the said building and other public members around the building, while upholding the Law.

6.5.1 Rules and Regulations for Class D Municipal Corporation

1. Applicability :

- The periodic structural inspection audit shall be necessary to all existing buildings **except,** detached houses, semi-detached houses, apartments which are used solely as places of residence, mix use occupancies less than G + 1 storey and temporary buildings.

- Periodic duration for structural audit : The periodic structural inspection audit shall be carried out on the following frequency :

  (a) After every **15 years** for buildings of detached houses, semi-detached houses, apartments which are used solely as places of residence, mix use.

  (b) After every **10 years** for all other buildings like institutional, commercial, hospital, assembly, etc. and buildings excluded as mentioned in (a).

2. Audit :

- Structural design inspection report of existing building having height more than 45 m prepared by the structural engineer and shall be checked by structural institution like Indian Institute of Technology Government college of engineering having specialized course in structural engineering.

  Note : IF the building owner/society do not complete the repairs/restorations, as per the directions in the Structural Audit Report within 6 month of submission of Audit Report, then they are also liable to be punished under section 471 and 352 (B) of the MMC Act.

6.6 BUDGET ESTIMATE

- Building maintenance is considered as a major activity in the construction industry because it is essential whether the buildings are large or small, simple or complex, located in urban or suburb. They must be well maintained to ensure their functionality and services during their life cycle.

- Budgeting and cost estimating are vital tools for project planning in the construction and play a important role in both preconstruction and construction phases of a project. While preparing budget estimate there will be always budget constraint, competing priorities in maintenance needs of any structures. The quality of maintenance activities is significantly influenced by the amount of budget allocated For the preparation of maintenance budget estimate following factors are considered a various maintenance types; and existing maintenance cost estimate. The main strategies for building maintenance are preventive strategy and corrective strategy. The quality of maintenance activities is significantly influenced by the amount of budget allocated.

The object of this budget estimation is :

1. To ascertain the necessary amount of money required by the owner to complete the proposed work.

2. To ascertain quantities of materials required in order to programme their timely procurement. To procure controlled materials, if any, like cement, steel, chemicals etc.; quantities of such materials are worked out from the estimate of work and attached with the applicant for verification.

3. To calculate the number of different categories of workers that are to be employed to complete the work within the scheduled time of completion.

4. To assess the requirements of tools, plants and equipments required to complete the work according to the programme.

5. To fix up the completion period from the volume of works involved in the estimate.

6. To draw up a construction schedule and programme and also to arrange the funds required according to the programming.

6.6.1 Maintenance Cost Planning and Estimating

- Maintenance cost includes all costs of keeping the building up to an acceptable standard. It relates to the direct cost of maintenance such as material, labours, equipment, scaffolding and tools as well as indirect costs such as administration, management and the inevitable overhead costs. When the demands of maintenance are identified, cost of the maintenance should be a prior estimate to measure resource availability and how much work should be scheduled in each period. Although cost estimates for building maintenance are normally prepared over the period to predict the likely cost of such works over the life of the buildings, they can be considered in a single annual maintenance programme.

- Determine the target cost limit for maintaining programme works. Inform setting the annualised maintenance budgets and available funding constraint. Provide cost information to assist decision makers to make informed decisions. Inform what asset investment are funded or not funded and then revise life cycle cost plan. Like any program or plan, maintenance budgets will be subjected to change and adjustment and it must be based on forecasting or predicting aiming to best utilise fixed maintenance resource to meet the fluctuating maintenance workload. Having a proper preventive maintenance strategy can reduce corrective maintenance cost, leading to reach optimal maintenance zone. The optimal zone is where the two costs are balanced. Once funds are approved for the maintenance budget, efficient use of this money requires wise internal allocation of the funding at the operational level or locating this optimal zone.

6.6.2 Maintenance Budget Estimates

- Before approval of any maintenance or adaptation work, an estimate of likely cost and time is needed. The maintenance manager prepares an estimate of cost and time to assess the cash needs, labour requirements, and priorities of jobs.

- Following are the **main purpose of budget estimate :**
    1. To make comparative evaluation between competitive estimates.
    2. To establish long and short-term budget requirements.
    3. To apportion or allocate money from an existing budget.
    4. To provide evidence for a request for extra money where necessary maintenance or adaptation is not covered by an existing budget.

- The estimator will require both data and an agreed system by which an estimated cost may be calculated. The data available will govern the accuracy of the estimate prepared. Preparing budget estimate will require following information for greater accuracy:
    (i)   Nature of proposed work.
    (ii)   Extent or scope of work.
    (iii) Method of operation.
    (iv)   Restrictions of any kind.
    (v)   Current labour costs and availability.
    (vi)   Past data of performance for similar jobs and conditions, usually obtained from feedback or historical records.
    (vii) Direct or contract labour consideration.
    (viii) Specialist services and consultancy fee.
    (ix)   Standards and level of specifications.

- These estimates can be detailed or approximate, depending on the purpose for which these are required. Approximate estimates are quick and simple method of assessing likely costs. These should be based on past records and performance which need to be adjusted taking into account any current inflation trends. For budget purpose cost estimation can also be predicted by using projected trends. Adjustments also need to be made for variation in condition Or nature of the work with regard to the past data.

- Following systems of cost estimation are commonly used for budget allocation for maintenance jobs:

(i) Cost per Plinth Area Covered :

- The plinth/floor area is usually measured gross inside the external walls where the maintenance jobs are carried out. This plinth area cost varies for variety of items of maintenance job and conditions under which the job is likely to be carried out.

(ii) Cost per Volume :

- This approach is based on cost per unit volume of building elements involved in maintenance job. This method also accounts for height. In a large size building, gross area of the external walls is multiplied by height. This approach is more suited to buildings, such as, hospitals, factories, and other buildings with large variation in heights. This approach also depends on the conditions and items of maintenance.

(iii) Cost per Element (Item-wise) :

- This method is based on the cost of each element (item wise) involved and related to an overall classification with reference to total floor area. Each element is broken down into the number of items to be considered for overall accuracy of maintenance estimate. Typical sub-elements or items can be, such as, Plaster, Painting, Polymer modified mortar, brick work, roof terrace waterproofing, structural components etc. This approach provides more accurate estimate for maintenance budget purposes.

Sr. No.	Particulars	Amount
1.	SECTION - I : MISCELLANEOUS ITEMS	8,000.00
2.	SECTION - II : SURFACE PREPARATION	97,000.00
3.	SECTION - III : CRACK / HONEYCOMB AREA REPAIR	7,700.00
4.	SECTION - IV : STRUCTURAL REPAIRS	163,500.00
5.	SECTION - V : MASONRY REPAIR	28,500.00
6.	SECTION - VI : CIVIL WORKS	309,000.00
7.	SECTION -VII : WATERPROOFING AND PROTECTIVE COATINGS	308,000.00
8.	SECTION - VIII : SANITARY & PLUMBING ITEMS	158,400.00
9.	SECTION - IX : PAINTING WORKS	290,000.00
	TOTAL	1,370,100.00
	CONTINGENCIES	68,505.00
	GRAND TOTAL	1,438,605.00
	SAY	1,439,000.00

6.6.3 Estimating for a Future Budget

- Estimate of cost is necessary for allocation of money to carry Out maintenance and adaptation work in the future. This requires an assessment of cost estimate for appropriate allocation of money in advance. Such an estimate can be carried out based on the trade or department, activities or the project as a whole. Information on nature and cost is collected from previous years and future cost is estimated by considering rate of inflation and other changes in conditions, contents and specifications. Two important components of any estimate are labour cost and material cost. Size of labour force required can be determined by using appropriate labour output under the specified conditions. It may be noted that labour output is likely to be smaller for piece meal jobs. Material costs can also be calculated by calculating quantities of materials allowing for material wastages, inflation and price trends of materials likely to be used for the maintenance.

- The amount of budget required can be based on the following costs :
  - (i) All wage rates of labour force including basic wages plus bonus or other statutory contributions and fringe benefits.
  - (ii) Estimated material costs considering price trends.
  - (iii) Maintenance establishment and overhead charges.
  - (iv) Any element of Profit required.
  - (v) All in costs to cover supervisory and other non-productive staff of maintenance wing.

The object of this budget estimation is :

1. To allocate budget and
2. To use it for comparative statement for tenders from outside contractors.

6.6.4 Estimating for Current Budget

- This is required for assessing the cost incurred on a Specific item of job from the total budget allocation. Sometimes this is required to compare with the predicted cost used in the preparation of the budget and previous feedback information obtained from the experts. This cost estimation may also be required for apportioning the budget for carrying out a specific job within certain allocated budget for the different trades, sections or departments. An estimate of how much money should be spent on any of these specific could be prepared and this would enable a certain control to be exercised over the current budget money allocation for unspecified maintenance works.

- The actual method of calculation will depend on the nature of wok and can range from spot item pricing to a unit rate pricing system based on actual quantities of labour and material required. A similar procedure may be adopted for requesting extra finance. Over and above the current budget allocation except that the reasons and cost would be very detailed to convince management and justify such extra expenses.

Important Points

- **Structural Audit :** Structural audit is "the inspection or examination of the building, to evaluate the strength so as to improve its appropriateness, safety, efficiency". This extent of damage or deterioration depends on the quality of work at the construction level. The deterioration of buildings can be the result of various factors including fire damage, frost action, chemical attack, steel erosion during the life span of the structure.

- **Distress Survey :** Distress survey is an examination of concrete for the purpose of identifying and defining area of distress. While it is referred in connection with survey of concrete and embedded reinforcement that is showing some degree of distress. Its application is recommended for all buildings and structures. The system is designed to be used for recording the history of the project from its inception to completion and subsequent life.

- **Visual Inspection :** Visual inspection of a structure is the most effective qualitative method of evaluation of structural soundness and identifying the typical distress symptoms together with the associated problems.

- **Audit :** Structural design inspection report of existing building having height more than 45 m shall be prepared by the structural engineer and shall be checked by structural institution like Indian Institute of Technology Government college of engineering having specialized course in structural engineering.

- **Budget Estimate :** Budgeting and cost estimating are vital tools for project planning in the construction and play an important role in both preconstruction and construction phases of a project. While preparing budget estimate there will be always budget constraint, competing priorities in maintenance needs of any structures

Questions for Practice

1. Define structural audit.
2. Explain the term distress survey.
3. Write step by step procedure of conducting structural audit.
4. Enlist the non-destructive test. And explain any one.
5. Explain visual inspection.
6. Write the necessity and importance of structural audit.
7. Write rules and regulations of structural audit.
8. State the format of structural audit report.
9. Explain the stages involved in structural audit.
10. Stages distress survey.
11. State and explain the term budget estimation.
12. Define budget estimate and state its importance.

❑❑❑

www.ingramcontent.com/pod-product-compliance
Lightning Source LLC
LaVergne TN
LVHW082118190726
843495LV00012B/1695

9 789389 825428